Principles and Practices
of
Unbiased Stereology

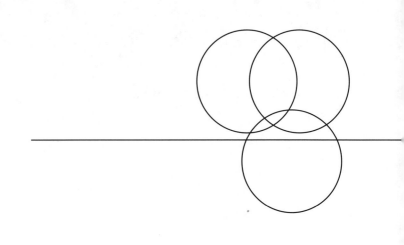

Principles and Practices of
Unbiased Stereology

An Introduction
for Bioscientists

Peter R. Mouton

The Johns Hopkins University Press

Baltimore and London

9 8 7 6 5 4 3 2 1

The Johns Hopkins University Press
2715 North Charles Street
Baltimore, Maryland 21218-4363
www.press.jhu.edu

Library of Congress Cataloging-in-Publication Data

Mouton, Peter R.
    Principles and practices of unbiased stereology : An introduction for
bioscientists / Peter R. Mouton.
        p.   cm.
    Includes bibliographical references (p.   ) and index.
    ISBN 0-8018-6797-5 (pbk. : alk. paper)
    1. Stereology.   2. Microstructure—Measurement.   I. Title.
Q175.M8777 2001
502′.8′2—dc21   2001000967

A catalog record for this book is available from the British Library

To my Sammie Lee, who knows what counts

Contents

Preface

W*hile current theories* and controversies surrounding modern stereology will be of interest to its practitioners, there are many professional biologists who want to become familiar with the principles and practices of this
technique, rather than its more arcane aspects. Currently their number far
exceeds the people who are available to teach them. Thus an introductory text
is needed to help disseminate information on the applications of the new, theoretically unbiased stereology to biological tissue. Those who wish to go on
to explore the scholarly debate and the theoretical discussion will find much
of interest in peer-reviewed articles in the *Journal of Microscopy* and *Acta Stereologica,* the proceedings of the International Society for Stereology, and
courses and workshops on stereology.

For those not drawn to the many theoretical sides of stereology, the good
news is that a comprehensive understanding of mathematics is not needed to
effectively apply new stereological methods to biological tissue. Biological scientists do not need an advanced understanding of its theory to design, super-
vise, and publish sound stereological studies. What they need is a good source
of tissue, the ability and resources to process and visualize the biological
objects of interest, and a commitment to obtaining reliable parameter estimates. With this foundation in place, most biologists are academically well
prepared to complete reliable stereological studies. After a decade of collab-
oration with biologists on stereology projects, I have learned that they do well
with a clear understanding of the terminology, a basic historical perspective,
and an understanding of what to do and what to avoid doing. A possible analogy is that of driving a car or using a computer; neither activity requires an
in-depth knowledge of what makes these machines work; otherwise car driv-
ers and computer users would be rare indeed. However, a bit of knowledge
beyond the most basic information improves performance and helps in solving problems as they arise.

This book is intended as a primer on the application of good stereological procedures to biological tissue. Thus I have consciously avoided advanced
theoretical and practical discussions in the interest of providing a broad

overview of the field. Persons interested in greater conceptual detail or more information on specific biological applications than I have provided here are encouraged to attend a stereology workshop and to consult the extensive citations in the bibliography.

In view of the wide diversity of biological organisms, and the wide variety of study designs possible to test specific hypotheses about them, it should not be surprising that there is no single approach, no particular computer software, and no "one size fits all" stereology design that can accommodate all studies and guarantee good results in all biological situations. Instead of following recipes, each investigator has the responsibility to develop and implement stereology designs that avoid bias in their particular application of theoretical stereological techniques to real-life biological issues. This approach ensures quality control and minimizes the inadvertent introduction of specious results into the scientific literature.

The purpose of this book is to provide readers with an understanding of the major concepts of the new stereology, particularly the most important sources of stereological and nonstereological bias that one encounters when attempting to quantify morphological structures in tissue sections. In making stereology less mysterious, my hope is that users will become comfortable enough with the new stereological procedures to complete their first study, and in time will find new applications and seek greater theoretical knowledge; this will naturally lend itself to more complex designs and analyses.

I wish to thank my stereology colleagues in the United States, including Professor Arun Gokhale, and my colleagues in Copenhagen and Arhus, particularly Professor Hans Gundersen, and Drs. Mark West, Bente Pakkenberg, and Arne Møller, for freely sharing their knowledge and friendship during my years studying stereology in Denmark and afterward. Of course, I accept full responsibility for the ideas expressed in this book; they are my own and are based on 15 years (and counting) of experience with the theory and practice of unbiased stereology. Finally, I sincerely hope the reader comes away with a better understanding of modern stereology after reading this book. If this happens, I will feel that the effort was worthwhile. Either way, I welcome comments and criticisms.

Principles and Practices
of
Unbiased Stereology

Introduction

Stereology describes the analysis of biological tissue in 3-D; "new" refers to the use of theoretically unbiased sampling and estimation methods to assess the first-order stereological parameters—volume, surface area, length, and number. The beauty of the new stereology is its clarity. When we get an estimate using unbiased methods, we can be confident that the estimate is reliable. If we make an error along the way, it is usually obvious and easily fixed before the estimate is completed. This is not the case for older, assumption-based stereology, also known as biased stereology. Estimates from biased stereology are based on faith in assumptions and models, an approach more appropriate to religion than to science.

In the past four decades, the field of stereology has produced theoretically unbiased approaches to estimating all four of the first-order stereological parameters. During this time there has been a rapid increase in the number of studies using stereological approaches (Figure I-1). When properly applied, these procedures allow users to make valid estimates of stereological

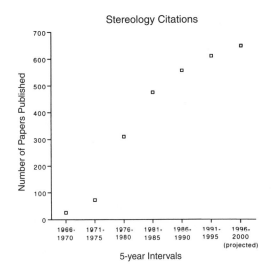

Figure I-1. Results of Medline search for publications with term *stereology* or *stereological* in title or abstract

parameters and their true variability. Yet despite the current popularity enjoyed by stereology, a significant amount of argument continues to swirl around the designation *unbiased*. This situation arises because the terms *biased* and *unbiased* have dual meanings, one mathematical and one colloquial. The mathematical definition is the one intended by stereologists. In an effort to defuse this debate, some stereologists have proposed using the terms *design-based* or *assumption and model free,* in reference to unbiased stereology. My personal feeling is that the "unbiasedness" of stereology applies strictly to theoretical situations. The term may not apply to all applications to biological tissues.

We can never know for sure, of course, how accurate our data are because we rarely, if ever, have results from gold-standard methods for comparison. For this reason the final arbiter of our accuracy must be how well we have succeeded in avoiding the numerous, well-known sources of potential bias introduced by the quantitative analysis of biological objects in tissue sections. Like many things in life, the best definition for good stereology is that you know it when you see it.

Good stereology includes sufficient sampling to capture most of the biological variation in a parameter. If we properly apply unbiased sampling and estimation techniques, but fail to adequately sample the tissue of interest, there is a risk of basing our conclusions on accurate but imprecise data. Another reason for thorough sampling is to reduce the effects of a single error. Stereological studies are done by humans, and we can all make an error in the best of conditions. With software it is trivial to eliminate the most common calculation errors; however, computerized systems still need their users to identify the biological features of interest in tissue sections. Stereological estimators make it possible to estimate millions or even trillions of objects by counting fewer than two hundred. However, with such a tool, an innocent identification error, in conjunction with inadequate sampling, can easily lead to an unreliable estimate.

On the other hand, sampling beyond the point of diminishing returns is not a productive use of time, effort, or material resources. As emphasized throughout this book, good stereology is also efficient sampling. To achieve optimal efficiency, we rely on the adage "Do more less well" to determine the level of sampling needed for stable, reliable estimates with a minimal level of effort.

Before discussing the design, implementation, and interpretation of good stereology studies, a more basic question arises: When is it worth the time, effort, and resources to analyze biological tissue using new stereology? Stereologists asked to consult on a project are often told by investigators that in a simple study with two or more treatment groups, the focus is on *relative* differences between groups for one or more stereological parameters. While these investigators see the point of estimating parameters for the entire population, such rigor is not necessary for routine studies. Instead, the thinking goes, the goal is to compare mean differences in a parameter between two groups of individuals. Therefore, the best method is the most efficient method available to obtain this estimate in a small sample from each group. In this way of thinking, as long as the data are collected by a person blind to the treatment groups, and the same methods are uniformly applied to all tissue analyzed, the final test of whether the study was done right is whether standard statistical methods show statistically significant results. If significant differences in groups are found, regardless of the methods used to estimate the parameters in the study, there is justification for publishing the results and moving on.

Notwithstanding the enormous pressures experienced by biomedical scientists to publish scientific results, with somber reflection one can appreciate the highly artificial distinction between absolute differences of population parameters and relative differences between sample estimates. Parameters at the population level are always the focus of a study; a sample estimate in a small number of individuals is simply a vehicle for making inferences about population parameters. Sound stereological procedures have been designed to avoid the factors that can distort sample estimates of biological objects in tissue sections. Yet if these procedures are modified by a variety of unmeasurable and nonverifiable assumptions for the sake of convenience, they will no longer support the investigator's expectation of accurate results. So the question remains: When should we use good stereological methods?

Perhaps the best justification for using good stereological procedures is what we intend for the results. Inferential statistics assume that dependent variables are correct; otherwise, inferences directed toward the population of interest will lead to statistical errors. When the primary goal is to apply the results to a defined population, then good stereology allows the results to be extrapolated beyond the small sample of individuals analyzed. Therefore, the ultimate justification for using good stereological approaches is to generate reliable parameter estimates in a small sample that can be extrapolated to the population parameter and that can be analyzed with confidence using powerful inferential statistics.

The price for the prestige that accompanies unbiased stereology is that its practitioners must understand the sources of potential bias, then expend the time and effort to avoid them. In some cases principal investigators have difficulty justifying the added cost and time when numerous "quick and dirty" approaches can generate morphometric data in less time and at less cost. However, it can be argued that despite the emphasis of unbiased stereology on increased rigor, good stereological approaches are not always the least efficient approaches available for morphological analysis of biological tissue. In addition to generating highly questionable results, assumption-based approaches with so-called shortcuts can in the long run easily exceed the time, effort, and resources required by good stereological procedures. Evidence of the inefficient methods available can be found in several examples in the morphology literature in which the total number of cells is determined by counting tens of thousands of cell profiles in a large number of tissue sections. After many hours have been spent counting these profiles, the results achieve the undesirable goal of methodologically biased and imprecise results. We know the results are inaccurate because the cells in question do not fit the models and assumptions of the assumption-based method.

Second, only a few cases can be analyzed by using this approach because the sampling is not efficient. Despite these heroic efforts, then, the next scientist to approach the same research question must face the reality that the previous results are not yet confirmed; the study must be repeated using theoretically unbiased geometric probes and systematic-random sampling. In short, despite the expenditure of a tremendous amount of time, effort, and resources, without the use of good stereological methods, little reliable information will be obtained with regard to the biological question of interest.

In contrast, if the study is done using good stereology, there is no need to repeat it. The job is done right the first time, at a reasonable cost and with an optimal expenditure of time and effort, and the results are reliable, defensible, and publishable.

Still, one cannot say categorically that good stereological methods are appropriate for all areas of scientific investigation involving morphological features of biological tissue. There are numerous qualitative end points that focus on biological parameters that are difficult to assess using stereological procedures, including differences in cell shape; the presence or absence of specific proteins; differences in staining intensity; or the presence or absence of particular genes, second messengers, molecular markers, etc.

In the opinion of this author, one criterion exists for deciding whether studies involving biological morphology should be quantified using good stereological procedures: *If the experimental conclusions are to be supported by statistical testing, then theoretically unbiased stereological methods should be used.*

1

The History of Stereology

This chapter presents a broad outline of the historical events leading to the current era of modern stereology. The term stereology *was first introduced in 1961, and after a brief period of reliance on assumption- and model-based classical geometry, moved into assumption- and model-free (theoretically unbiased) approaches to quantify 3-D objects of biological interest. The chronology of these developments provides insight into the rationale for modern stereological approaches.*

The term **stereology,** which has its origin in the Greek word for solid, *stereos,* was introduced into the scientific vernacular in the early 1960s. In 1961 a diverse group of biologists, geologists, and materials scientists met in the Black Forest of Germany to discuss problems associated with the quantification of 3-D objects from their appearance on 2-D sections. At that meeting, Professor Hans Elias, a biologist from Germany, suggested that *stereology* was a convenient term to describe the subject of their discussions, and the term was introduced to describe a discipline focused on analyzing the structural parameters of objects based on their appearance on 2-D sections.

Early in the course of a series of professional meetings, which were attended by scientists from diverse disciplines and backgrounds, it was realized that stereologists had been discovering and using the same ideas for decades. Until the 1960s, however, there had been no multidisciplinary forum for discussions about issues related to stereology.

In 1962 the first International Congress for Stereology was held in Vienna, Austria. The formation of the International Society for Stereology (ISS) followed, with Professor Elias as the founding president. The first goal of the ISS was to focus its collective resources on a deceptively simple question: *Can reliable quantitative information about 3-D structure be obtained from 2-D tissue sections?*

For the next four decades scientists interested in stereology from diverse academic institutions, government agencies, and private industries presented their work at the ISS's biennial meetings. Today researchers in the natural sciences and all basic science divisions in medical universities, as well as in departments of geology, mathematics, and engineering, are using stereology to answer research questions.

The growing popularity of stereology in biomedical research in recent years follows recognition of the importance of morphological changes in testing hypotheses about biological phenomena. In addition to the emergence of stereology, the 1960s witnessed the culmination of a century of technological advances in microscopy that led to the visualization of biological structures with unprecedented clarity. The development of highly sensitive immunological probes for the visualization of specific proteins and cells and increasingly higher resolution of specific tissue constituents using sophisticated and powerful microscopes led to the inevitable questions related to stereology: How many biological entities are there? What is their size and length? Is there a true difference in volumes and surface areas? To help answer these questions, biologists and medical scientists began turning to their colleagues with an interest in stereology.

From Geometry to Stereology

People living over 6000 years ago in Egypt were the first to quantify the basic stereological parameters: volume, surface area, length, and number. Surface area was of particular interest to these civilizations. Each year after the great Nile overflowed its banks, the resulting deposits of soil obliterated boundaries between plots of land. After the floodwaters receded, farmers needed to be able to reliably reestablish the boundaries of their property. This was a job for the rope handlers, who had a highly practiced ability to take steps of exactly equal size; they marked off property boundaries with straight lines and sharp corners using great lengths of rope. These primitive surveyors brought order and stability to early human civilizations, and in doing so were among the first humans to use geometry to resolve practical problems.

By 4000 B.C. the rope handlers developed methods for aligning stones with the movement of the sun, the moon, and other heavenly bodies. Using the same principles of 3-D triangulation as satellite-based global positioning systems, the rope handlers' techniques addressed questions based on the spatial orientation of objects relative to each other. In time, Egyptian architects became interested in objects that occupied volumes, the massive pyramids. By 1000 B.C. these architectural projects attracted the attention of nearby Greek city-states, where there was plenty of stone, and slave labor was controlled by powerful kings. Greek kings sent ambassadors to Egypt to learn the secrets of *geometry* (Gk, "to measure or survey the earth").

One of the most famous early Greeks to visit Egypt was Pythagoras (582–500 B.C.), who eventually returned to Athens and established a school dedicated to the study of geometry as the key for understanding truth. Today he is best known for the Pythagorean theorem, perhaps one of the most recognizable concepts of classical geometry: *In any right triangle, the square on the hypotenuse equals the sum of the squares on the other two sides.*

Pythagoras developed this theory after observing numerous practical applications of triangulation in Egypt, as exemplified by the following practical application:

From the top of a wall 60 feet high, a rope is stretched to the bottom of a fence 50 feet from the base of the wall. What is the length of the rope?

Solution. The rope forms the hypotenuse *c* of a right triangle; the height of the wall is side *a,* and side *b* is the distance from the bottom of the fence to the base of the wall.

Let c = the length of the rope

Then $c^2 = a^2 + b^2 = 60^2 + 50^2 = 3600 + 2500 = 6100$

$c = \sqrt{6100} = 78.1$ feet

Though the Egyptians invented geometry, the Greeks were the first people to broadly apply geometric principles. They used them in places of worship, offices, gardens, amphitheaters, gymnasiums, roads, wagons, and sailing vessels. In return, by the fourth century B.C. geometry had made a major contribution toward establishing Greece as the architectural, philosophical, and academic center of the civilized world. A major contributor to the use of this geometry was Euclid (330–275 B.C.).

Classical Euclidean geometry provides a set of tools for constructing objects and structures and for understanding mathematical relationships. Unfortunately for biologists, however, the same approaches cannot be applied to biological structures, which do not fit into the idealized molds of classically shaped objects. The extent to which biological entities vary from classical models depends on two factors: the actual geometric features of the objects and the degree of variation between objects in the population of interest. In contrast to classical geometry where the dimensions of objects can be defined, biological objects vary within and between groups. It is precisely the arbitrary and varying nature of biological objects that introduces bias when classical formulas are applied to such objects. Despite the severe and obvious limitations of classical geometry, prior to the advent of unbiased stereology in the 1980s, the primary methods for quantifying biological objects in tissue sections were assumption- and model-based approaches using classical geometry.

From the Renaissance to the Year 2000

Until Europeans rediscovered classical Greek mathematics during the Renaissance, we have no evidence of efforts to develop theoretically unbiased stereological methods. From the fifteenth century to the present, however, a number of contributions established the theoretical foundation for modern stereology.

• In 1635, Buonaventura Cavalieri showed that the mean volume of a population of nonclassically shaped objects can be estimated in a theoretically unbiased manner from the sum of areas cut through the objects.

Cavalieri's method allows for the morphological estimation of the total volume of any population of objects from the area on a systematic-random sampling of sections through the objects.

• In 1777, the naturalist George Leclerc Buffon presented the Royal Academy of Sciences in Paris with the needle problem.

Buffon had a strong interest in the laws of probability. He showed that a needle tossed at random onto a grid of lines intersects each line with a probability directly proportional to the length of the needle, with no further assumptions. Buffon's principle provided the theoretical basis for estimating the total length and the total surface area of nonclassically shaped objects.

• In 1847, the French mining engineer and geologist Auguste Delesse showed that the profile area of a random section through a population of objects is proportional to the expected value of the object's volume.

Delesse was interested in quantifying the various phases of mineral deposits in rocks. (A phase is a homogeneous fraction of physical material that can be separated from the rest.) Specifically, Delesse wished to quantify the relationship between the phase area on the cut surface of a rock and the total phase volume in the entire rock. By comparing the area of phases on cut surfaces with the total phase volume in the whole specimen, he showed that the total area of a phase on each cut surface is proportional to the total phase in the entire specimen. Today, the Delesse principle provides the basis for estimating the volume of nonclassically shaped objects based on their profile areas on random sections.

• In 1925 Swedish mathematician S. D. Wicksell described the corpuscle problem to explain why accurate estimates of the number of biological objects with arbitrary, nonclassical sizes and shapes cannot be obtained from profile counts on single, thin histological sections.

Wicksell wanted to know the number of thyroid globules in the thyroid gland. From sections cut through the thyroid and carefully reconstructed in 3-D, he showed that the number of globules in a given volume of tissue, the number per unit of volume, or the density, N_v, cannot be estimated from the number of globule profiles on the cut surface of sections, N_A. This became know as the "corpuscle problem." Since then, numerous attempts have failed to overcome the corpuscle problem by using a "correction formula," which simply adds further bias through assumptions and models that do not apply to biological objects.

• In 1984 D. C. Sterio published the disector principle, the first unbiased method for estimating the true number of objects in a given volume of tissue, N_v.

D. C. Sterio is the pseudonym of a well-known Danish stereologist who, for personal reasons, wishes to remain anonymous. However, careful readers will note the similarity in the letters in disector and D. C. Sterio. The disector method was designed to overcome the corpuscle problem without assumptions, models, or correction factors. The method Sterio developed uses two virtual planes, a disector pair, and two conventions introduced by Professor Hans Juergen Gundersen in the late 1970s: an unbiased counting frame and unbiased counting rules. D. C. Sterio's disector principle provided the first theoretically unbiased method to sample and count the true number of biological objects per unit of volume, without assumptions about their size, shape, and orientation, in a defined region of tissue.

With the advent of the disector principle, stereologists realized that quantitative methods of morphological analysis could in theory overcome the most severe forms of bias introduced by cutting 3-D objects into 2-D sections. By the year 2000, unbiased stereological methods were developed for making accurate, precise, and efficient estimates of first-order stereological parameters and their variability. Before beginning the task of analyzing the four first-order stereological parameters, we continue this chapter with a discussion of the unique geometry of biological objects.

The Geometry of Objects in 3-D

All objects can be said to belong to one of two groups: natural or man-made. A second universal observation is that whether objects are naturally occurring or man-made, all by definition exist in three dimensions. Man-made objects are characterized by classical geometric shapes, including planar flat surfaces, squares, circles, half-circles, spheres, curves, etc. Take a moment to notice the unique morphological properties of these objects. Note how these shapes change from one region within the object to the next. Man-made objects have surfaces that are flat, with long, straight edges and that change direction sharply or in generally uniform, rounded curves. Consider this book, for example: a 3-D object with flat, planar surfaces of approximately equal dimension and thickness, with sharp, uniform edges in a symmetrical orientation around a central binding. Based on its morphological features, we intuitively know the object did not evolve naturally. The lines on each page are evenly spaced, with uniform margins at the top, bottom, and sides. Except for differences in the exact words on each page, there is little variation from one page to the next. If you had two copies of this book, you would find few differences between them. For the remarkable consistency of books we can thank the printing press, first developed in the fifteenth century and improved for efficiency ever since. For the majority of man-made objects, the industrial revolution in the twentieth century provided manufacturing processes for producing a desired number of objects of the same size, shape, and consistency according to a predesigned blueprint based on classical geometric shapes.

Naturally Occurring Objects in 3-D

Contrast the morphological features of man-made objects with those of naturally occurring objects that constitute organic, living creatures, such as plants and animals formed by biological processes, and inorganic substances such as minerals and rocks.[1] Although there are a few notable exceptions, especially among the crystalline structures of inorganic objects, an obvious characteristic of both organic and inorganic substances is the general lack of morphological uniformity within and between objects. That is, naturally formed objects show relatively high variability in morphological features within the same individual, and among individuals in the same population. Populations of trees in a forest, mice of a particular species, and rocks in a valley floor all differ from other individuals within their group. We refer to this variation as natural variation, which can be the result of natural selection and in certain cases artificial selection. Techniques of artificial selection well known to crop

---

1. Although this book focuses on biological objects, the category of naturally occurring objects includes inorganic specimens with morphological features that vary predominantly as a result of geophysical factors.

farmers, geneticists, and animal breeders allow humans to change natural variation by selectively breeding for specific morphological features. As bio-scientists, we expect natural variation in the morphological parameters of biological objects. As stereologists, we seek to quantify that variation and in the process to test hypotheses concerning the functioning of tissue that is normal, injured, genetically altered, diseased, etc.

Biological objects exist in the 3-D spaces of tissue. The biological objects themselves occupy some of this space, which is defined by the dimensions of first-order stereological parameters. Cell *volumes* occur in 3-D spaces, membrane *surface areas* in 2-D spaces, and the *length* of linear elements in 1-D space; the *number* of objects has no dimension (0-D). These dimensions are further defined by their units, for example, cubic millimeter, square millimeter, and millimeter, and no units, respectively. The decision of which parameter(s) should be quantified in a particular study requires a thorough examination of the purpose of the research, especially the hypothesis of interest.

Suppose we are interested in quantifying the morphological basis for impaired gas exchange in a mouse model for pulmonary hypertension. The primary function of lung alveoli is the exchange of gases between blood and the outside environment. Therefore the surface area of alveoli, rather than their total number, is the first-order stereological parameter with the most direct relationship to the study. The exchange of gases is the function, the membrane is the associated morphological structure, and the surface area is the first-order stereological parameter of interest. In another example, if neuronal innervation in a defined brain region is the focus of a study, then the parameter of primary interest is the total length of neuronal fibers in the target region. In this example, innervation is the function, the nerve fiber is the associated morphological structure, and length is the first-order stereological parameter.

Although a variety of structures and parameters could be quantified, the goal is to identify the objects and parameters with the closest relationship to the physiological function of interest. Identification of the parameter of interest is the first step in the development of an efficient, unbiased design. Before we delve further into the specifics of unbiased study design, however, the next section emphasizes the importance of avoiding assumptions and models when quantifying morphological parameters of biological objects.

Since the invention of the microscope in the late 1500s, biological objects have been viewed according to their appearance on 2-D sections. An important question is what morphological perturbations are caused by cutting 3-D objects into 2-D sections? The process of preparing tissue for microscopic analysis using light, confocal, or electron microscopy can also affect the structure of a biological object since it requires treatment of the tissue with a variety of processes and reagents, many of which can alter the morphological appearance of the tissue and the objects contained within it. Because biological tissue is generally softer than the knives used to cut sections, it is frequently embedded in a relatively hard matrix such as paraffin or plastic, or the tissue is frozen into a hard block prior to sectioning. Using a sharp knife (a microtome), tissue for microscopic analysis is cut into sections that are thin enough to allow light to pass through them, allowing us to view the objects as magnified images. Between the time sections are cut and objects and tissues are visualized, however, the tissue must first be stained to reveal particular populations of biologically interesting objects. As you will read many times throughout this book, it is important to keep in mind that microscopic study of tissues requires procedures and reactions that can change the morphological features of tissues in significant ways. Therefore, the goal for good stereology is to quantify 3-D parameters using a combination of 2-D sections and 3-D thinking.

If the objects of interest in biological tissue were classically shaped as spheres, cubes, and tetrahedrons, the classical geometric formulas taught for centuries would be appropriate for calculation of first-order stereological parameters. If cells were spheres, we could easily measure their mean volume using the classical geometric formula $V = 4\pi\, r^3/3$. As alluded to earlier, cells are not spheres, and biological objects rarely, if ever, meet the stringent criteria required for use of the classical geometric formula. Furthermore, biological objects by their nature vary from one individual to the next, in contrast to classical models. Defining a biological object according to a model fails to include fine variation in size, shape, or orientation.

As we will discover in later chapters, once biological entities and parameters are identified for the functions of research interest, the next step is selecting an unbiased geometric probe that contains the correct number of dimensions. That is, a probe with the dimensions that, when summed with the dimensions of the parameter of interest, equal or exceed 3, the total number of dimensions in biological tissue. For example, to avoid stereological bias for estimation of total length (1-D) for linear objects in tissue sections, the tissue must be sampled with a probe that is at least 2-D, such as a plane or the surface of a sphere. This approach ensures that the objects of interest in the tissue cannot "hide" within the 3-D space of the tissue, and that the parameter of interest will be quantified with a probability that is directly related to its magnitude.

To quantify populations of biological objects, good stereology requires that we avoid assumptions and models that could, under any circumstances

or treatments or natural variation, introduce bias into estimates of stereological parameters and their fine variations. Although this goal sounds somewhat ambitious at first, the task is made easier by the assumption- and model-
free approaches of stochastic geometry and probability theory.

Stochastic Geometry

The term *stochastic* is used to describe random processes, objects, or variables. In biological systems we find stochastic processes and objects every-
where, beginning with an individual's genetic code, its DNA. Variation in
DNA arises out of random combinations of parental DNA, followed by
countless physiological reactions that lead to a phenotype, or physical expression of a particular genetic code (a genotype). Because populations of individuals differ in their DNA, we expect to observe morphological differences
in phenotypes, from outward physical characteristics to cellular differences
in stereological parameters. With identical twins, who share the same genotype, the expression of these genes results in striking similarities: the same
sex, the same blood type, the same hair and eye color. They also have the same
nose, ear lobe, and lip shapes. Identical twins experience similar speech and
social development and usually vary in IQ by fewer than five points. Despite
these similarities, identical twins exhibit definite physical and personality differences that are obvious to family, friends, and careful observers. Thus, even
in the most extreme example of individuals following the same model, underlying stochastic processes lead to biological variability.

Because of stochastic processes, we expect more or less variability in the
morphological parameters of individual organisms. And because of this expected variability we find that as a rule, classical geometry does not apply to
biological objects. In later chapters we discuss the probes and sampling designs of unbiased stereology that are based on stochastic geometry and probability theory. The goal of these approaches is to make reliable estimates of
stereological parameters, and in doing so to capture the inherent variability
of a parameter in the population of interest.

Summary

Modern stereology arose in response to critical weaknesses in the applications of powerful assumption-based geometric formulas to problems surrounding the morphology of naturally occurring 3-D objects. Because biological objects as a rule do not fit neatly into classical geometric spaces,
model- and assumption-based formulas introduce systematic error (bias)

that once present, cannot be removed by correction factors. Instead of using assumptions, models, and correction formulas, the goal of modern stereology is to avoid all known sources of bias in the quantification of naturally occurring objects. This goal has been successfully achieved using a combination of stochastic geometry and probability theory that constitutes the theoretical basis for modern stereology.

2

Bias

This chapter discusses the need to avoid assumptions and models that can add systematic error (bias) to estimates of biological structures. Stereological bias causes estimates made from samples to diverge from expected values for the distribution of the parameter. An important feature is that once it is present, stereological bias cannot be quantified, corrected, or removed. Understanding how to identify and avoid sources of stereological bias is the first step toward making theoretically unbiased estimates of structural parameters in biological tissue.

In **Chapter 1** we reviewed the fundamental differences between naturally occurring and man-made objects and showed that different methods are needed to quantify populations of objects in each group. In this chapter the importance of avoiding stereological bias is reviewed, starting with the formal beginning of modern stereology in the 1960s and the extension of its applications as it made use of the experience of geologists, engineers, and mathematicians.

Forty Years of Modern Stereology

The First Decade

The first decade of the modern era of stereology coincided with a dramatic growth in microscopy techniques and immunologically based staining methods. These developments allowed biologists to visualize specific populations of cells and other structures, which led to increasing interest in quantitative techniques. In 1971 *Science* published a review by Professor Hans Elias that was intended to draw attention to the stereological knowledge gained at ISS conferences during the previous decade. An important point of this review was the growing need for biologists to understand the large inaccuracy that occurred when 3-D parameters were quantified based on their 2-D appearance on tissue sections. Elias's review noted that misinterpretation of flat images on single sections gives rise to "perpetuated errors" in the scientific literature. Flat 2-D images, he pointed out, provide erronenous representations of the full dimensionality of a structure as it exists in 3-D. He noted that these errors grew from assumptions that biological objects have classical shapes and that each part of the structure in 3-D has an unequal probability of being sampled by a sectioning knife. Elias noted, therefore, that accurate estimates of biological objects can never be found using model- and assumption-based geometry. In spite of intentions to sample and estimate biological structures, Elias argued, unless one paid serious attention to their nonclassical shape, the morphometry of biological objects would always be misleading.

For many biologists interested in morphometric analysis of biological structure (i.e., measurement of phenotype form), Elias's warning coincided with the end of the first decade of modern stereology. For the next generation of stereologists, the first decade had sent a clear message: The future of stereology was assumption and model free. The decade can be summed up as follows:

- Biologists discover methods used in materials sciences
- Weakness in model-based stereological methods realized
- Improvements in tissue processing and visualization methods

The Second Decade

The paradigm shift from assumption-based to assumption-free methods did not occur without resistance from biologists who were new to the "new stereology." Nevertheless, the development of theoretically unbiased methods grew rapidly from 1971 to 1981. During this decade the ISS held five international meetings and increasingly focused on two journals, the *Journal of Microscopy* and *Acta Stereologica*. These publications and their support by ISS members provided the growing community of stereologists with an international forum for peer-reviewed communications.

The shift from classical geometry toward stochastic geometry and probability theory was a turning point in the development of modern stereology. During the 1970s, mathematicians, geologists, and engineers from the materials sciences brought to the ISS their particular expertise and experience in analyzing nonclassically shaped 3-D objects. After considering the problem from a variety of viewpoints, mathematicians in the ISS identified stochastic geometry and probability theory as the disciplines most likely to establish a strong basis for the theoretically unbiased analysis of stereological parameters for nonclassical objects. By the start of the third decade of modern stereology, the idea began to germinate that these disciplines could be used effectively to make theoretically unbiased estimates of the four first-order stereo-logical parameters. Moreover, it was realized that if assumption- and model-free estimates of biologically interesting parameters could be made without introducing appreciable artifacts from tissue processing, then the variation observed in these estimates would strongly reflect biological variability. In the meantime, new and theoretically unbiased sampling designs were developed for estimating parameters of biological interest. (See the papers by G. Math-eron, R. Miles, and L. Cruz-Orive in the bibliography.) The second decade then, was one of seeking mathematical justification and solidifying stereo-logical theory on the basis of probability and stochastic geometry.

The Third Decade

In the 1980s biologists involved in the development of stereological meth-ods began to clearly envision approaches that if properly applied, could be used to obtain reliable estimates of population-based morphological param-eters by analyzing tissue from a small sample of randomly sampled individ-uals. To make this idea a reality, however, enough tissue would have to be sam-pled to avoid introducing a high degree of sampling error; second, processing of biological tissue would have to avoid introducing artifacts. This meant that for the new stereology to be effective, new geometric probes would have to be combined with new, highly efficient sampling strategies, and new proce-dures would be needed to avoid processing artifacts. By the early 1980s, stere-ologists had the theoretical foundation to estimate population parameters of biological interest. However, if the new stereology was to have practical value beyond the theoretical stage, it was imperative for biologists to provide the proof of concept.

The third decade witnessed two critically important breakthroughs in the application of modern stereological approaches to biological tissue. First, stereologists were successful in developing and refining the disector princi-ple, the first geometric probe for the theoretically unbiased analysis of the total number of objects in biological tissue. As indicated earlier, this break-through was made possible by a novel geometric probe, the disector, devel-oped by D. C. Sterio in 1984. When the disector principle was used in com-bination with an unbiased counting frame and unbiased counting rules, the number of intersections between the 3-D disector probe and the objects of interest in the tissue was equal only to the true number of objects in a known volume of tissue, without further assumptions. Because this probe provided theoretically unbiased results, it meant that estimates for a total number could be made in a small sample and extrapolated to the expected value at the population level.

The second breakthrough, the application of systematic-random sampling to biologically important reference spaces by Gundersen and Jensen in 1987, showed that theoretically unbiased probes such as the disector could be efficiently applied to biological tissue without introducing stereological bias. This latter breakthrough confirmed that for most biological tissue, the adage to do more, less well will ensure adequate within-individual sampling while conserving the time, effort, and resources spent on reducing the major contribution to variation in a sample estimate—biological variation. Rather than oversampling under the false assumption that more sampling equals a better estimate, this strategy applies sampling efforts where they will produce the greatest return. When combined with the new geometric probes, it provides an efficiency that is on a par with previous assumption-based (biased) stereological approaches. Thus the third decade of modern stereology was characterized by the development of techniques for obtaining efficient, accurate, and precise estimates of parameters in biological populations.

The Fourth Decade

The fourth decade of modern stereology has come to a close. As alluded to in the introduction, the rise of new stereology as the method of choice for analyzing biological morphology has been accompanied by resistance among practitioners of the old stereology. One major objection was the term *biased*, which, as indicated earlier, has both colloquial and mathematical meanings. In colloquial usage, "biased" implies intentional misrepresentation; it was this meaning that led some to misinterpret new stereology as a reproach of earlier methods. Understandably, established scientists did not appreciate the term *biased* used in this context in reference to their work. The proponents of the new stereology countered that mathematically speaking, the term *unbiased* refers to a theoretical method to estimate the expected value of a parameter, that is, the weighted average of the parameter for the population. In response to this confusion over semantics, theoretically unbiased stereological methods began to appear in the literature under a variety of aliases: *design-based, assumption and model free,* and *the new stereology.* Regardless of the name, the same ideas apply: They describe methods designed to make reliable inferences about an expected distribution of a parameter based on a relatively small number of measurements in a relatively small sample of nonclassically shaped biological objects in a defined reference space.

A second problem was that the new stereological approaches were perceived as radical. They did not follow the time-honored mechanism for adding knowledge to the biological sciences, namely, building on existing precedents in the literature. This was because a multidisciplinary, multinational coalition of scientists had never before focused its energy and talents on a nonmilitary goal: quantifying objects in 3-D based on their appearance on

2-D sections. To achieve this goal, a paradigm shift was necessary, which meant changing the rules that had developed over many decades. As a result, the new stereology required more, not less, time; more careful sampling of tissue; and greater labor in morphometric analysis.

To many opponents of the new stereology, these facts ran counter to the idea of progress. They argued that biological research needed faster, cheaper, and less labor-intensive methods. Perhaps the most salient criticism of the new stereology was the dogmatic approach of leaders in the field, stereologists who approached the subject with what can best be described as religious fervor. To be accurate, these stereologists argued, no shortcuts could be permitted; assumptions could not be allowed. Notwithstanding the criticisms, some deserved and some not, by the end of the fourth decade of modern stereology, the shift in favor of unbiased approaches was well under way. The rest of this chapter examines some of the issues and problems that stereologists face as the discipline continues to refine its techniques and theory.

Sources of Bias: Sampling and Estimation

Consider a hypothetical survey funded by the Ford Motor Company to assess the automobile preferences of registered car owners in the United States. Before investing resources in designing new cars, the company needs to understand the car-buying preferences of American car owners. The accuracy of this information will depend to a significant extent on how the survey is designed and carried out. One approach would be to poll Ford workers outside their Detroit production plant. However, basing survey information on a sample of the target population that will be affected by the results could provide misleading interpretations. Obtaining an accurate and therefore theoretically unbiased assessment of the true preferences of American car buyers requires effort, time, and thought to avoid selecting a biased sample for the survey. A second factor that will strongly affect the results of such a survey is the way the questions are posed. For instance, one could say: "Fords are the best cars made in America, don't you think?" Although this question posed in Detroit would most likely generate results in favor of the company, the results would provide little information on the true thinking of American car buyers. A more theoretically unbiased question, posed to a more representative sample of the American population, might be: "If you were going to purchase a car today, what type of car would it be?" Surveys, like stereological studies, require both the sample and the probe to be theoretically unbiased.

Similar considerations apply when designing stereological studies to assess biological tissue. The design should generate reliable estimates about the magnitude and variation—the mean and standard deviation—of structural parameters of interest and do so efficiently. Efficiency counts because, like the cost for sampling car owners in the hypothetical car survey, maximizing the benefits of the resources used is an important consideration for bioscientists. In both cases, the goal is to capture most of the variation in a parameter by efficiently sampling a small but representative percentage of the total population.

The Reference Space

For parameter estimates to have meaning at the population level, that is, for sample estimates to provide theoretically unbiased estimates of parameter distributions, we must confine our estimates to bounded reference spaces that can be consistently sampled from one individual to the next in the population of interest. Here the term *reference space* refers to the anatomical region where the biological objects of interest are located within the tissue. For example, if we are interested in counting the cells in the cerebral cortex of the rat brain, then the rat cerebral cortex is our reference space. The characteristics of a reference space that are needed in order to make theoretically unbiased estimates of population parameters are the following:

- It must be unambiguously defined
- It must contain the tissue of interest
- It must be 100% available

For both theoretical and practical purposes, these characteristics must be met by all reference spaces from one individual to the next.

If the reference space lacks clear boundaries with other anatomical regions, then the parameter analyzed does not have a clear distribution in the population. On the practical side, an inconsistent reference space in which the borders of the anatomical region change from one individual to the next will have little meaning at the level of the population. Imagine attempting to count the total number of people in a busy street during rush hour traffic. Unless the space is bounded, the number of objects in the space will be constantly changing, as would our estimates at any given time.

Finally, a reference space must be available for sampling. To make reliable estimates about first-order parameters in an anatomical region, the entire region must be available for sampling. Being available does not imply that the entire reference space must be analyzed, but rather that all parts of the reference space must be available to contribute to the estimate of the parameter. If part of the reference space is missing and not available to contribute to the estimate, then the estimate will not be accurate for the entire reference space; in this case, the estimate would apply to an ill-defined and nonreproducible distribution of the parameter.

Stereological versus Nonstereological Bias

Bias can be defined as stereological or nonstereological. In stereology, accuracy refers to how closely a sample estimate converges on the expected value after increased sampling. It is a mathematical term that has a definite meaning in terms of the expected value of a parameter. If a method produces an estimate that deviates from the expected value, that estimate is inaccurate by the amount of the deviation. Accuracy can be considered the bull's-eye of a target. The expected value of a population parameter is the center of the bull's-eye. Deviation of a mean sample estimate from center can be considered inaccurate.

Expected value refers to the expected value of a parameter for the entire population of interest. However, the true or expected value of a parameter for an entire population is usually unknown. If this value were known, then we would hardly be justified in spending time, effort, and resources on making estimates. Because biomedical scientists require accurate methods that are also efficient, new stereology was developed as an efficient means to estimate the expected value of a parameter. It is important to realize that methods can be accurate or inaccurate independently of the application of the method to a particular tissue.

Systematic error or bias consists of an unknown but consistent amount of deviation (inaccuracy) in an unknown but consistent direction. When systematic bias occurs because of assumptions and models related to the objects of interest, the bias is stereological; once it is present, stereological bias can never be removed because it is embedded in the theory of the methodology. It is based on assumptions, models, or correction factors that cannot be verified and whose degree of error is not known. Systematic error from other causes, that is, nonstereological bias, *can* be removed, if the source of the deviation can be identified and eliminated; this type of removable bias is known as uncertainty.

An example of stereological bias is the systematic error introduced by insufficient dimensions in the probe and the parameter. *The sum of dimensions in the probe and the parameter must equal at least 3.*

The reason for 3 is quite simple and requires no mathematical proof. To avoid stereological bias, the total number of available dimensions in tissue (i.e., 3-D) must not exceed the sum of dimensions in the parameter and the probe. For example, a pre-1990s approach to counting cell number was to count the number of profiles on a tissue section. In this case the sum of di-

Table 2–1 Parameters, Probes, and Their Dimensions (Dim)

Parameter	Dim.	Structure	Probe `	Dim.	Sum of Dimensions
Volume	3	Volume	Point	0	3
Area	2	Surface	Line	1	3
Length	1	Linear	Plane	2	3
Number	0	Cardinality	Disector	3	3

mensions in the parameter (number = 0-D) and the probe (tissue section = 2-D) is less than 3. The stereological bias arises from the fact that cells may occupy the remaining dimension in the tissue and that these cells have a reduced probability of being counted. Table 2-1 shows the requirements for dimensions in parameters and probes to avoid this form of stereological bias.

Accurate results depend not only on selecting a theoretically unbiased method but also on avoiding the multitude of potential sources of nonstereological bias inherent in the preparation of tissue for morphological analysis. These sources include artifacts of the fixation, preservation, embedding, sectioning, staining, etc., required in order to observe biological objects.

Other nonstereological sources of error range from ascertainment bias, which occurs when estimates based on samples from one population are extrapolated to another population, to failure of stains to penetrate through tissue and fully reveal the objects of biological interest (recognition bias). For this reason it is important for scientists designing sampling and estimation procedures to review potential sources of nonstereological bias that can cause systematic error. Although the list of potential sources of nonstereological bias is beyond the scope of this book, several of the most common sources are reviewed in Chapter 9.

In summary, bias can result from both stereological and nonstereological sources as shown below:

Stereological Bias
• Faulty "correction" factors
• A sum of probe and parameter that is less than 3
• Incorrect models or assumptions

Nonstereological Bias
• Incomplete staining
• Ascertainment bias
• Improper calibration or observer bias
• Incorrect mathematics

The outcome of all sources of bias is the introduction of error (variation) that cannot be accounted for by biological variation or sampling error.

Error (Variation)

⇓

Systematic error (bias) + Nonsystematic error (unbiased)

⇓	⇓
Stereological bias (inaccuracy)	Biological variation (theoretically unbiased)
+	+
Nonstereological bias (uncertainty)	Sampling error (theoretically unbiased)

Precision and Bias

Biological Variation

The expected value of a parameter varies from one individual to the next, resulting in biological variation. Biological variation is defined at the level of the population and therefore is rarely known and cannot be easily found. However, by excluding all known sources of stereological and nonstereological bias, we can make reliable estimates of biological variation from our sample estimates.

For the purpose of adding and subtracting variation in mathematical equations, biological variation (BV) is squared and expressed as biological variance (BV^2)—the mean-squared difference between the estimate of the parameter and its expected value. The sum of relative variance (CE^2) and biological variance (BV^2) constitutes the total observed variance (CV^2) in a sample estimate (this is discussed in more detail in Chapter 10).

total observed variance $= CV^2 = BV^2 + CE^2$

Because CV^2 and CE^2 can be estimated from sample estimates, by rearranging the above formula and solving for BV^2 we are able to assess the contribution of biological variation to the variation in sample estimates. This source of variation can be decreased by sampling more individuals, or increased by sampling fewer individuals from the population. As shown in the section on sampling error, the formula permits us to optimize our sampling design for maximum efficiency.

In contrast to a single measurement using a highly calibrated instrument, stereology estimates consist of a group of values that aggregate around a central tendency. The amount of spread around the central tendency is dependent on how many individuals are analyzed, that is, how much effort is used to capture the biological variation in the parameter of interest. Thus, biological variation is the true "noise" around the expected central tendency of the population. Like the example of surveying registered car owners, if we analyze a representative, random sample of individuals in a population, the variation observed in the sample values can be representative of the expected biological variation for the parameter in the population. Biological variability can only be reduced by increased sampling, by including more values from individuals in the population of interest; this in turn will capture a larger amount of variability.

A common misconception is that biased results from assumption- and model-based methods can be overcome through brute force—a higher level of sampling. Increasing the sampling level leads to more precise results, but not necessarily to more accurate results. As shown later, higher levels of sampling with a biased method fail to accomplish this goal; in fact, increased sampling with a biased method causes the central tendency of the sample results to deviate more, not less, from the expected central tendency for the population. As mentioned earlier, once stereological bias is present, it cannot be removed through any means, including increased sampling.

Sampling Error

The second source of random variation observed in sample estimates is sampling error—variation arising from the intensity of sampling within each individual. It appears as "noise" or sampling error around each data point. Sampling error is expressed in terms of the coefficient of error (CE), which for mathematical equations is expressed as relative variance, CE^2. In relative terms, reducing sampling error costs less than reducing biological variation.

The degree of sampling noise for each estimate depends on two factors: the number of sections analyzed (between-section variation) and the number of regions analyzed within each section (within-section variation). The total of all of factors related to sampling noise (i.e., the total within-subject variation) can be modified by increasing or decreasing the intensity of sampling. Higher levels of within-subject sampling reduce sampling noise, while lower sampling levels increase it. By partitioning the CE into the sampling error contributed by between-section and within-section sources, one can efficiently increase the precision of individual sample estimates by sampling more sections or sampling more within each section.

To be theoretically unbiased, the sampling approach within each reference space must be random. Random sampling guarantees that all parts of a reference space will have an equal probability of being sampled. An important outcome is that the full biological variability of a parameter within a reference space is captured in the sample estimate. As discussed further in Chapter 10, the goal of an efficient stereological design is to capture most of the biological variation with a minimum of sampling noise.

Pilot Studies

The goal of an efficient stereology design is also to capture most of the biological variation in a population parameter by lightly analyzing a minimum number of individuals. To achieve this goal, we use an approach called "do more less well," which avoids unnecessary oversampling within each individual and focuses on reducing biological variation in the sample estimate. This approach requires a pilot study in a few individuals sampled at random from the population. As detailed in Chapter 10, the sampling error is set at a known level, then the sampling design is optimized for maximum efficiency based on partitioning the observed variation into biological variation and sampling error. The optimal level of sampling is the intensity of sampling within each individual that achieves a substantial reduction in the observed variation per hour spent analyzing tissue. When the point of diminishing returns is realized, that is, when further increases in sampling intensity achieve only minor reductions in the observed variation, then resources are best spent analyzing more individuals sampled at random from the population.

Biased versus Unbiased Approaches

With the concepts of accuracy and precision in mind, we can elaborate on the distinction between theoretically biased and unbiased approaches (see Figure 2-1): *In theoretically unbiased approaches, increased sampling converges the sample estimate on the expected value of a population parameter.*

A theoretically unbiased method generates a sample estimate that provides a theoretically accurate estimate of a population parameter. As mentioned earlier, one feature of an unbiased method is that with increased sampling, the sample values will converge on the expected value; with a biased method, further sampling will cause the sample value to converge some unknown distance away from the expected value. When bias is present, the expected value of the sample estimate will deviate systematically from the expected value of the parameter. Statistically speaking,

$$\text{bias} = E(X) - \mu$$

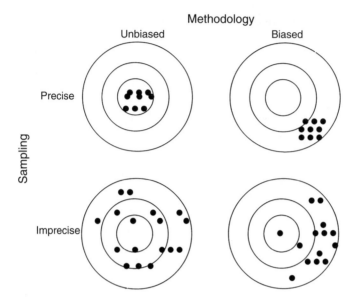

Figure 2-1. Expected results for unbiased/biased stereology methods and precise/imprecise sampling results

where bias is the systematic error, $E(X)$ is the expected value for an estimate of parameter X, and μ is the expected value for parameter X.

Variance in statistical terms refers to random error, the mean-squared difference between an estimate of parameter X and the expected value, $E(X)$. An important characteristic of sample variance is that it is unrelated to bias, accuracy, or mean-square error. Second, variance, in contrast to bias, can be measured empirically from sample data. Like parameters of biological interest, the variation for a given parameter can be estimated in a theoretically unbiased manner through the use of theoretically unbiased geometric probes and theoretically unbiased sampling. Through careful experimental design and good stereology, parameters and their variance can be estimated for a population of biological organisms.

Correction Factors

A close examination of classical geometry reveals a number of attractive formulas which, if applied to biological objects, would provide highly efficient but highly assumption- and model-based approaches for estimating biological parameters on tissue sections. Over the years, many individuals have proposed a variety of correction factors in an effort to "fit" classical formulas into biological applications. The problem with this approach is that correction formulas themselves invariably require assumptions and models that are rarely, if ever, true for biological objects. These formulas fail to "correct"; instead, they add further assumption- and model-based systematic errors. They fail because the magnitude and direction of bias are not known. For example, if we assume that a population of cells is on average 35% nonspherical, and if this assumption is not correct, then correcting raw data by this amount would lead to biased results.

Although correction factors may be correct in theory, this is only true when their assumptions are met, which for biological objects is rarely the case. Verifying the assumptions of correction factors is so time and labor intensive that it is prohibitive for routine studies. How does one quantify the non-sphericity of a cell? Furthermore, how does one take into consideration the variability in nonsphericity of a population of individuals, much less a population of cells? Or in the case of a study with two or more treatment groups, the differences in nonsphericity between groups receiving different treatments? Clearly one is on a slippery slope when attempting to quantify biological objects using formulas based on assumptions, models, and correction factors. A dozen or so correction factors have been proposed to remove stereological bias arising from the application of classical geometry to biological objects, but in good stereology the only appropriate use of such factors is in those narrow cases where the assumptions of the correction factor can be clearly defined and verified. In most cases, however, correction formulas with nonverifiable assumptions sound too good to be true, and usually are.

The Reference Trap

Bias can also be introduced by assuming that a reference space is constant. This kind of bias has earned its own name—the reference trap. It is discussed more fully in Chapter 5, but here an example shows how seriously it can affect results.

In the middle 1950s, numerous studies of cell numbers in the brains of young and old cases at autopsy reported significant losses of brain cells per year after age 50. This result provided a compelling explanation for the reduction in memory and other cognitive functions shown to occur during aging. Because brain cells are responsible for learning and memory, the age-related loss of brain cells provided a clear explanation for age-related loss of learning and memory skills.

This conclusion led to studies of brain aging by the German anatomist and stereologist, Herbert Haug. Haug wondered if the process of fixation to preserve the tissue caused differential shrinkage of brains from individuals of different ages. His careful studies showed that neural tissue undergoes fixation-induced shrinkage that is *inversely proportional* to the age of the tissue;

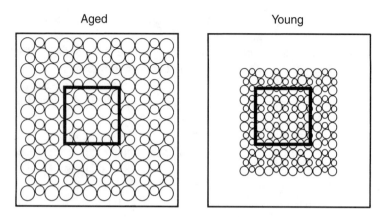

Figure 2-2. An example of potential bias associated with density estimators in aging studies. The shrinkage in reference volume is greater in young brain tissue than in that from older brains following equivalent periods of aldehyde fixation, leading to decreased density in older brains *in the absence of true cell loss.*

after comparable periods of fixation, older tissues shrink substantially less than young tissue (see the 1984 paper by Haug et al. in the bibliography). These results suggested that differential shrinkage of the reference volume, not brain cell loss, may explain the lower densities of brain cells in random sections from old brains (Figure 2-2). Rather than brain cell loss, this age-related difference in densities induced by fixation may result from decreased shrinkage of the reference space secondary to increased sclerosis (hardening) of fatty material (myelin) with age.

Summary

In the past decade and a half, the development of theoretically unbiased principles for analyzing biological structure and its variability have revolutionized the way we analyze the morphological parameters of biological tissue. In the 1990s biologists in a wide variety of subdisciplines effectively applied new stereological approaches to estimate biologically interesting parameters. The advent of commercially available computer-assisted stereology systems has dramatically increased efficiency while reducing the resources needed to complete stereological studies. Even with computer-assisted stereology, however, in the words of the mathematician-stereologist Professor Luis Cruz-Orive of Spain, unbiased stereology is "a committed task." An efficient stereology design avoids all known sources of experimental bias, both stereological and nonstereological, and allows optimization of data collection for maximum efficiency.

3

Sampling

Theoretically unbiased systematic-random sampling of biological tissue for stereology is discussed in this chapter. Random sampling guarantees that the sampling is theoretically unbiased; systematic sampling ensures that the sampling is efficient. Systematic-random sampling begins with a clear definition of a reference space, the three-dimensional unit of tissue that contains the biological objects of interest. Once a reference space has been unambiguously defined, the second step is to sample the space in an unbiased, efficient manner. Systematic sampling is the most efficient strategy to capture the greater part of variation in the reference space in the shortest amount of time. Thus, systematic-random sampling ensures that all objects of interest in all parts of the reference space are efficiently sampled in a theoretically unbiased manner.

Population Parameter Distributions

Recall the four morphological parameters of interest, also known as first-order stereological parameters: volume, surface area, length, and number. Now consider a particular population of naturally occurring objects, for example, all the livers in laboratory mice of a particular age, gender, and strain. If we pick a particular parameter of interest in this population, such as total number of liver cells, we can begin to envision a population parameter of interest: for instance the mean number of liver cells in female C57BL/6J mice aged 3 to 6 months. Of course, we do not expect that all female mice in this age group will have the same number of liver cells; rather, we expect the true number of liver cells to vary around a central tendency. This variation is referred to as *biological variation* and is set by a variety of forces, including natural selection, artificial breeding practices (artificial selection), and environmental forces (diet, toxins, etc.). The range of values for a given parameter for all of the individuals in the population is called the parameter distribution. For the example given here, the parameter distribution covers all 3- to 6-month-old, female, C57BL/6J mice (x-axis) and the value for the total number of liver cells for each mouse (y-axis). From this parameter distribution we can find the central tendency (mean) and variance for the population parameter of interest.

Recreating an entire parameter distribution for a routine stereological study is inefficient, impractical, and unnecessary. What we can do instead, is take a small number of individuals at random from the population of interest, sample their tissue, and estimate the morphological parameters of interest in a theoretically unbiased manner. If we have effectively avoided the major forms of stereological and nonstereological bias while estimating the parameters, the resulting estimates will provide estimates that are both theoretically and actually unbiased for the target population parameter.

Applications of Stereology to Biological Objects

Stereological information frequently provides the basis for testing specific hypotheses about biological systems. Essentially every field of biomedical research at one time or another focuses on changes in one of the four first-order stereological parameters. Terms such as *degeneration, toxicity, atrophy/hypertrophy, dysgenesis,* and *proliferation* all refer to alterations in one or more of these parameters. Stereological approaches also provide the ability to estimate a second-order stereological parameter: variability. Together, stereological parameters and their variability contain the information necessary to test a multitude of hypotheses involving putative changes in biological morphology (Table 3-1).

Anatomists use stereological parameters to test hypotheses concerning evolutionary, interspecies, and functional differences for a variety of independent variables, for example, drugs, experimental manipulations, or changes in habitat. Biomedical scientists assess stereological parameters in tissue to understand the underlying causes of disease. Behaviorists quantify stereological parameters in relation to functional changes in cognitive, motor, and sensory systems, so-called brain–behavior connections. Immunologists, pharmacologists, and biochemists use stereology to understand the morphological mechanisms underlying physiological responses in tissues. Molecular bi-

Table 3–1 Stereological Approaches Can Be Used to Test Hypotheses

Study	Morphology	Hypothesis
Comparative anatomy	Vertebrate wing	Evolution
Young vs. aged	Brain tissue	Aging/development process
Disease vs. normal	Bone marrow	Indicators of cancer
Mutation vs. wild-type	Liver cell	Genetic origins of disease

Table 3–2 Changes in Structural Parameters Can Lead to Disease

Structural parameter	Change	Pathology	Disease condition
Volume	Increased myocardial volume	Myocardial hypertrophy	Cardiac failure
Number	Increased cell number	Cellular proliferation	Cancer
Length	Reduced bone length	Stunted growth	Down syndrome
Area	Decreased alveolar surface area	Impaired gas exchange	Pulmonary fibrosis

ologists and geneticists make stereological estimates of the phenotypic characteristics of particular genotypes. Changes in structural (stereological) parameters are also common indicators of disease, as shown by the examples in Table 3-2.

Prior to the modern era of stereology, morphological parameters were studied using experts, computerized image analysis, and assumption-based stereological approaches, as discussed in the next section.

Morphometry Experts

The ancient Greeks established descriptive, qualitative observations as the basis for understanding biological processes, beginning with the scientific writings of Aristotle (384–322 B.C.). Later, Galen (A.D. 130–201), the royal physician at the School of the Gladiators in Rome, recorded some of the earliest qualitative descriptions of pathological changes in biological structure.

The invention of the microscope in 1590 led to a tremendous expansion of descriptive studies of biological tissue. The mass production of high-quality objective lens by the Carl Zeiss company in the early twentieth century stimulated technological improvements focused on greater magnification and higher resolution of microscopic objects. During the latter half of the twentieth century, the introduction of immunologically based techniques for staining specific proteins contributed to an explosion of detailed information about biological structures.

In biomedical research, expert proficiency lies in the ability to recognize and interpret qualitative changes in morphological parameters, based on the appearance of biological structures in gross and microscopic tissue. Expert morphological knowledge is built through training and experience, usually under the mentorship of other recognized experts. For the past two millennia, qualitative opinions of experts have formed the bulk of biological and medical knowledge.

With the increased ability to visualize biological tissue, however, came the realization of its extraordinary complexity, and the recognition of severe limitations inherent in expert-based assessments of biological tissue. First, expert knowledge is highly focused on a particular tissue. Almost as many experts are required as there are tissues in the body. An expert on blood cell precursors in the bone marrow of leukemia patients could not be expected to be expert in the analysis of liver biopsies of alcoholics, the degree of basement membrane thickening in the diabetic kidney, or the extent of brain degeneration in Alzheimer's disease. Nevertheless, by the mid-1950s there was great demand for the few biologists and medical specialists recognized as bona fide experts in their particular fields. Although some experts showed remarkably high throughput (sections per unit of time) and high intra-rater reliability (the same result on blind evaluations of the same material by the same person), by the 1960s it was clear that expert-based morphometry showed poor reproducibility between observers. This low inter-rater reliability meant that one needed strong faith in the opinion of one expert. Not surprisingly in the absence of a "gold standard," expert opinions became the source of heated controversies and arguments, which usually had no hope of being resolved.

In response, biomedical scientists began to rely increasingly on semi-quantitative terms for differentiating morphological changes, for example, none, mild, moderate, or severe; and, +1, +2, +3 . . . . To achieve greater rigor by avoiding observer bias, scientists also began to use observers who did not know the group identification for the tissue under study (blind observers). By the late twentieth century, a further limitation of expert-based morphometry emerged: Data that are based on semiquantitative results are not amenable to powerful parametric methods of statistical analysis; instead one must rely on less rigorous nonparametric approaches to test whether inferences are likely to be true on the basis of probability. The next step was to increase throughput of blind observers by taking maximum advantage of powerful computer technology in assessing biological tissue.

Computerized Image Analysis

The "computer-as-expert" approach to biological quantification was developed in response to the demand for greater objectivity in obtaining morphometric data. The hope was that applying computer-based technology to the analysis of biological tissue could revolutionize the collection of morphometric data. The major costs would be minimal: a computer, preparation of tissue, and a data collector blind to the identity of the tissue being analyzed. By the mid- to late 1980s, however, a consensus was emerging from studies using computerized image analysis systems that the accuracy of such data depended on how the software was programmed. Because computer programmers typically did not work closely with biologists, early programs were based on a number of assumptions and models related to classical geometric shapes, rather than on the morphologically diverse objects found in biological tissue. To achieve greater throughput, early computer-based attempts were made at pattern recognition, although these ultimately failed. Even today, the most advanced computerized approaches available still make less accurate assessments of biological tissue than a trained human.

A second problem with these semiquantitative computerized approaches is the extensive time required for editing out unwanted information from the data. By the early 1980s, the boom of computerized tissue analysis was over, once again sending the majority of bioscientists in search of experts. It was at this time that many biologists discovered that the new experts in morphometry were into something completely different.

Tools of the New Stereology

What biologists found was that morphometry now had available to it powerful new tools that allowed researchers to sample morphological variations in biological structures efficiently and precisely. These sample estimates could then be used to obtain accurate estimates of changes within a population by making a relatively small number of measurements in a few randomly sampled individuals. The following sections describe the sampling theory and the results that can be expected. First, however, the reader should become familiar with the basic terms of sampling methodology.

Sampling Terminology

The terminology of stereology includes a mixture of terms from statistical analysis and sampling theory, as the following summary shows.

Parameter: Population value being estimated in a sample
Sample: Individuals analyzed from the population
Estimator: Probe used to estimate a parameter
Estimate: Parameter from an estimator in a random sample
Reference space: Bounded region containing objects of interest
Expected value: True or expected value of a parameter for the population

Specific units in 3-D are associated with biological parameters, for example, total volume in cubic centimeters, cubic millimeters, cubic micrometers (Table 3-3). In addition to the terms borrowed from statistics, sampling uses some terms with specific meaning in stereology. These are briefly defined here.

Population: A well-defined set of individuals. Examples are all adult

Table 3–3 First-Order Sterological Parameters, Dimensions, and Units
Associated with Biological Reference Spaces

Structure	Parameter	Dimensions	Units
Tissue	Volume (V)	3	Volume (cm^3, mm^3, μm^3)
Surface	Area (A)	2	Area (cm^2, mm^2, μm^2)
Linear	Length (L)	1	Length (cm, mm μm)
Objects	Number (N)	0	—

Americans, the cells in a rat liver, or brains of 6-month-old mice of a specific strain.

Parameter: A well-defined numerical quantity relating to the population. Examples are the mean height of all adult Americans, the mean total number of cells in female rat liver, or the mean total brain volume of 6-month-old C57BL/6J mice.

Sample: A set of *n* individuals from the population. Examples are John Smith and Mary Doe ($n = 2$), twenty sections ($n = 20$) from the liver of female rat No. 25, or brains from five ($n = 5$) mice.

Population parameter: The true or expected mean value of a parameter in the population, $E(X)$.

Sample estimate: The value of a parameter estimated in a random sample of individuals from the population, $E(x)$.

Theoretically unbiased estimate: Zero difference between the expected value of a mean sample estimate and the expected value of the mean parameter for the population; $E(X) - E(x) = 0$.

Biased estimate: A nonzero difference between the expected value for a sample estimate and the expected value of the parameter $[E(X)]$ for the population; $E(X) - E(x) \neq 0$.

Theoretically unbiased sampling design: A mechanism that allows every unit in a defined population to have an equal probability of being included in the sample.

First-order stereological parameter: Volume (3-D), surface area (2-D), length (1-D), number (0-D).

Second-order stereological parameters: Variance, for example, coefficient of variance (CV^2), sampling error (variance) (CE^2), biological variance (BV^2).

Reference Spaces

As indicated earlier, the basic unit of sampling in stereology is the reference space. In general, a reference space is a 3-D form containing the objects

Table 3–4 Examples of Biological Reference Space
and Associated Function

Organ	Reference space	Biological function
Brain	Cerebral cortex	Cognitive activity
Heart	Left ventricle	Pump blood to arterial circulation
Lung	Alveolar sacs	Site O_2:CO_2 exchange

or tissue of interest. In biological studies, the reference space is an anatomically defined tissue that controls a physiologically important biological function (Table 3-4). The boundaries of a reference space are not determined by any stereological rule, but by the tissue properties of interest to the investigator. The analogous term from the older assumption-based stereology literature is "region of interest"; however, as discussed later, there are a number of important distinctions between a reference space and a region of interest (ROI).

The first step in a stereological study is to identify the correct reference space for the physiological function of interest. In most cases, the reference space can be deduced by an examination of the hypothesis being tested. For instance, to study the total number of cells in a brain region controlling cognitive function in mammals, one reference space could be the cerebral cortex. The reference space for a study of inhalant toxicity in rats could be the alveolar sacs of the rat lung. For a study testing possible drug nephrotoxicity in mice, the reference space could be a defined subregion of the mouse kidney containing nephrons.

We say that the reference space is deterministic, i.e., the borders of the reference space *determine* the value of the estimate. Variation in the criteria for definition of the reference space will produce variation in the estimate that is unrelated to the parameter of interest. If two investigators are interested in estimating a particular parameter, and the first investigator defines a relatively large reference space, the two investigators will generate different estimates for the same parameter. Yet, both studies are designed to estimate the expected or true value of the parameter. For both investigators to make unbiased and comparable estimates of the same parameter, both must sample within a consistently defined reference space.

The same argument holds for one investigator estimating a parameter in different individuals from the population of interest. The goal is to estimate the expected parameter for the population by making sample estimates in a small, random sample of the population. If a consistent reference space cannot be defined in each individual, then some variation in the resulting mean value will be attributable to unclear inter-individual definitions of the reference space, rather than variation in the parameter of interest. Thus, estimation of population parameters for a biological object or region assumes that the borders of the reference space are defined. The best way to achieve a defined reference space is to identify, sample, and estimate within an anatomical space that is naturally bounded. If the borders of the reference space are not clearly defined, the target distribution will be similarly undefined. For this reason it is meaningless to make sample estimates of a reference space with borders that cannot be clearly defined. In cases where a reference space cannot be readily defined, investigators should consult anatomical texts or seek collaborations with anatomists before attempting to use theoretically unbiased methods to estimate parameters.

In practical terms, the reference space is simply the region to be sampled in a particular study. The crucial prerequisites for a reference space are that it must be unambiguously defined and consistent, bounded, and entirely available for sampling. Making sample estimates only in sharply defined reference spaces focuses the study on a parameter distribution at the level of the population. When one attempts to carry out statistical testing (e.g., analysis of variance, ANOVA), an unknown and unmeasurable fraction of the variation in the sample estimate will be attributable to the poorly defined, inconsistent reference space rather than to true biological variation.

Because of the wide variety of possible reference spaces, it is not possible to list the criteria for all biological reference spaces. A few examples can be given for illustrative purposes, however. In the simplest case, for studies of the functions of an organ, the boundary of the reference space is the organ itself. For instance, the reference space for the total volume of the brain is the brain; for the total surface area of the lung it is the lung; and for the total number of liver cells it is the liver. It is clear that sampling and estimating parameters in such reference spaces can be easily related to a parameter distribution at the population level.

At the other extreme are poorly defined reference spaces, such as the functional subregions of the cerebral cortex. Although it is well known that these subregions are distinct in terms of the functions they control, with a few notable exceptions it is generally not possible to define each of them on anatomical grounds. To obtain theoretically unbiased estimates of stereological parameters, the anatomical tissue of interest must be distinctly delineated from adjacent regions; the delineated tissue must have functional significance; and it must be possible to clearly recognize the delineated tissue in all individuals. As in the case of the cerebral cortex, it is more difficult to define consistent, natural boundaries in subregions of larger organs than for an entire organ. In these cases the investigator must use anatomical borders with adjacent structures, staining characteristics, and other reliable criteria to identify consistent borders for the reference space before sampling.

Once a reference space has been defined, it is crucial that the full extent of that space be available for sampling. Again, this is because sample estimates of stereological parameters and their variation must relate to a defined parameter distribution. If part of the reference space is missing and therefore not available for sampling, then this tissue will not be included in the sample estimate. In this case, the sample estimate will deviate from the expected value of the parameter.

The borders of a reference space are deterministic: They determine the magnitude of sample estimates.

A poorly defined reference space introduces an unknown quantity of variation in sample estimates that makes it difficult to separate the total variation into variation from sampling error and that from biological variation (Chapter 10). For this reason it is imperative to make strict, biologically meaningful definitions of a reference space in relation to adjacent biological structures, and for the entire reference space to be available for sampling.

Region of Interest versus Reference Space

Prior to the advent of new stereology, many investigators used digitization and techniques of image analysis to quantify biological images. The software for this technology introduced the concept of region of interest (ROI). In image analysis, a digitized image of picture elements (pixels) provides the basis for selecting the ROI for further analysis. The purpose of defining an ROI is to select a population of pixels for analysis of the image; however, in contrast to a reference space, the population of pixels in an ROI is not necessarily well defined and reproducible. There is no requirement for consistent borders or for including all or part of an anatomical region. An ROI can be defined according to arbitrary criteria, for example, convenience, interesting subareas, availability of tissue, or personal preferences. Because the purpose of an ROI is simply to select pixels for analysis, it is not the same concept as a reference space. However, an ROI can be a useful reference space, provided it is defined according to anatomical boundaries that are unambiguous and consistent from one individual to the next.

There is a second important distinction between reference spaces and ROIs. Digitization procedures identify ROIs according to levels of transmitted light, which are set to arbitrary gray levels without regard to the functional or anatomical definitions of the objects. Digitization typically includes a wide variety of cells of different types, blood vessels, etc., which may not be a part of the tissue of interest, but which can easily be included in the analysis. Before the data can be analyzed using image analysis procedures, digital images must be first edited to ensure that only objects of interest are included in the ROI. When carried out properly, this editing process can be time-consuming and tedious and a source of bias when ignored.

A third caveat for quantitative methods based on an ROI is that light transmitted through tissue sections may be systematically altered by variations in staining intensity and section thickness. The resulting estimate will vary according to the extent that these factors vary. Diseased, aged, or drug-treated tissue may shrink more or less, or stain with more or less intensity, in comparison with normal tissue following equal tissue-processing procedures. Unfortunately these factors vary in an unpredictable and inconsistent manner; otherwise we could develop assumption-free mathematical approaches to minimize their impact on the variation of sample estimates. Morphological estimates based on sampling of an entire reference space avoid these sources of bias; estimates based on an arbitrary ROI may not.

The Power of Sampling

Using systematic-random sampling, biological stereologists have successfully estimated parameters in biological tissue ranging from a few thousand cells to a few trillion objects. For instance, there are about 6 billion humans on Earth, each possessing one brain, with each brain covered by a single cerebral cortex containing about 20 to 25 billion brain cells. Thus, we can estimate that the total number of brain cells on Earth is about 1.2×10^{11} ($6 \times 20 \times 10^9 = 120,000,000,000$). Though such calculations are obviously rough, for practical purposes measurement of the true number is both impossible and unnecessary.

Rather than making exact measurements with a highly calibrated instrument, new stereology focuses on theoretically unbiased estimates of population parameters, which in turn depend on efficient, theoretically unbiased systematic-random sampling and the use of assumption- and model-free geometric probes. This chapter describes sampling methodology and Chapter 4 discusses geometric probes.

Sampling with Equal Probability Equals Random Sampling

Randomness is an important consideration for all probability-based outcomes. Rolling dice, flipping coins, playing poker, and picking lottery winners involve probability-dependent outcomes. For the various outcomes to have their intended probabilities, dice must be shaken, cards shuffled, coins tossed in the air, and numbers selected at random. Deviations from these randomization procedures introduce bias by increasing the probability that particular outcomes will be favored over others. A basic tenet of theoretically un-

Table 3–5 Sampling Based on Equal Probability

Individual	Weight X (g)	Probability (P)	$P \times X$(g)
Flopsy	400	1/4	100
Mopsy	540	1/4	135
Cottontail	500	1/4	125
Peter	450	1/4	112

biased sampling is that all objects in the reference space must have an equal probability of being sampled (see Table 3-5) as illustrated in the following trivial example:

Goal: To make an estimate of the mean weight of four individuals.
Approach: Put four individuals in a hat. Sample each individual with equal probability and measure its weight. The sampling is carried out by placing four individuals in a hat and sampling each one, one at a time. After each individual is sampled, its weight is recorded. The question is: Before sampling begins, what is our prediction for the expected mean weight of all four individuals? All of the individuals have different weights; otherwise they have an equal probability of being sampled. Therefore the expected value of the sample estimate, $E(x)$, is the sum of the products of individual probabilities and the weight of each individual: $\Sigma (P \times X)$, mathematically expressed as follows:

$$E(x) = \Sigma(P \times X) = 100 + 135 + 125 + 112 = 472 \text{ g}$$

This example describes a theoretically unbiased sampling design: All objects of interest have the same probability of being sampled. Note that $E(X)$ is equal to the arithmetic mean for the population, μ:

$$E(X) = \mu = \Sigma X / n = (400 + 540 + 500 + 450) / 4 = 472 \text{ g}$$

Based on the assumption that all objects have an equal probability of being sampled, the expected mean weight of an object is 472 g. Now consider the following biased sampling design.

Imagine a hat containing four rabbits and that the ear lengths for each rabbit are different. These variable ear lengths introduce an unequal sampling probability in the experiment, ranging from the rabbit with the longest ears, Flopsy, with a 1/2 sampling probability, to the rabbits with the shortest ears, Cottontail and Peter, each with a 1/8th sampling probability. When we repeat the sampling to assess the mean weight of rabbits in the hat, we would expect the results shown in Table 3-6:

Table 3–6 Sampling Based on Unequal Probability

Individual	Weight X (g)	Probability (P)	$P \times X$(g)
Flopsy	400	1/2	200.0
Mopsy	540	1/4	135.0
Cottontail	500	1/8	62.5
Peter	450	1/8	56.3

As before, we can predict the expected value for mean weight as the sum of the products of each sampling probability (P) and its actual weight (X).

$$E(x) = 200 + 135 + 62.5 + 56.25 = 453.75 = 454 \text{ g}$$

Recall that the true mean weight and the expected mean weight were both 472 g based on an equal sampling probability. With an unequal sampling probability, the expected mean weight is now 454 g. Thus, before the experiment begins we predict that the mean sample estimate will differ from the expected value by -18 g. This difference between the expected mean sample estimate and the expected true value of the parameter is bias or systematic error.

In this simple example we know the differences in sampling probabilities and can predict the extent of bias. In biological tissue, however, we are entirely ignorant of the unequal sampling probabilities created by our sampling design. A variety of sources can introduce sampling biases that simultaneously over- and underestimate the true expected parameter. Factors that can contribute to sampling bias include:

Oversampling of "interesting" structures
Unequal sampling of the reference space
Undersampling of "uninteresting" structures
Artifacts of sectioning (e.g., lost caps)
Unequal sampling of individuals from population
Poorly defined reference space
Poor staining, incomplete penetration, artifacts, etc.
Biased geometric probes

Types of Random Sampling

As indicated earlier, random sampling is applied to biological structures at all levels of the sampling hierarchy, beginning with random selection of individuals from the population and proceeding from high to low levels of magnification within the reference space. The sampling hierarchy for biological tissues is shown in Table 3-7.

Table 3–7 Sampling Hierarchy for Biological Tissues

Level	Sampling	Magnification	Example
Individual	Independent-random	Low	Human, monkey, mouse
Tissue	Systematic-random	Low	Brain, heart, lung
Slab/section	Systematic-random	Low	Cortex, vessel, epidermis
Probe	Systematic-random	Low/high	Points, lines, planes, volume

Independent-Random Sampling

To apply results from a small sample of individuals to a complete population using, for example, inferential statistics, individuals must be sampled in a random *and* an independent manner. An example of nonindependent sampling would be to select individual 1 from the population, analyze a reference space in this individual, then select individual 2 based on the analysis for individual 1. Continuing in this manner, individual 2 is analyzed, and based on this result individual 3 is selected, and so on. This nonindependent sampling design would generate a biased sample, that is, one that is not representative of the target population.

By definition, independent-random means that the method is random—for a number series from 1 to 10, all numbers from 1 to 10 have an equal probability—and it is independent—samples are unrelated to each other.

Random versus Nonrandom Sampling

The following exercise demonstrates the differences between random and nonrandom sampling in selecting three objects from a population of ten objects.

Goal: To sample three objects from a population of ten objects.

O   O   O   O   O   O   O   O   O   O

Nonrandom approach: Select the first three objects from the population (sampled individuals shown as **O**):

O   **O**   **O**   O   O   O   O   O   O   O

In this example, all of the sampling units do not have an equal sampling probability—Os to the left of the series have a higher sampling probability than Os to the right of the series, leading to a biased sample. For humans, sampling with random numbers without help is difficult because the tendency is to pick some number with a higher frequency than others (e.g., 3, 7). The better alternative is to use computers and hand-held calculators with programs to generate numbers that are effectively random. These numbers can be printed in a table (random number table) for use in the stereology laboratory. An example of five computer-generated random numbers in the interval 1 to 10 from a random number table is 8, 1, 5, 7, 4. This random series of numbers can be used to make an independent-random sample of n = three objects from a population of ten objects.

> *Goal:* To obtain an independent-random sample of three objects from a population containing ten objects.

> *Independent-random approach:* Select three random numbers: 8, 1, 5. Then sample objects corresponding to these random numbers.

O	O	O	O	*O*	O	O	*O*	O
1	2	3	4	*5*	6	7	*8*	9

Once a number from a random number table has been used, it should be crossed out to avoid reusing the same numbers repeatedly. When all the numbers in the table have been depleted, a new set of random numbers should be generated.

This example illustrates a basic principle of theoretically unbiased, random sampling. In addition to being random, the above approach illustrates *independent* sampling. The selection of one object has no effect on the sampling of other objects. For independent-random sampling, each sampling unit has an equal and independent probability of being sampled. In the 1970s, stereologists expanded this idea into an equally unbiased but more efficient approach for sampling a population of objects; this is dependent-random sampling, also known as systematic-random sampling.

Systematic-Random Sampling

With systematic-random sampling, the first object is sampled at random, then each successive object is sampled systematically. When a uniform interval is used between successive sampling units, the term *systematic-uniform-random* may be used. Because this approach is almost exclusively used in sectioning biological material, the shorthand term *systematic-random* is most common. The systematic-random-uniform approach can be illustrated as follows:

> *Goal:* To obtain a systematic-random sample of three objects from a population of ten objects.

> *Systematic-random approach:* Select a proportion to sample to give the desired sample size; 10/3 = 3. Choose a nonzero random number be-

tween 1 and 3, for example, 2. Sample the object at that position, then sample every third object thereafter. The computer-generated random number = 2.

O   O   O   O   *O*   O   O   *O*   O   O
1   2   3   4   5   6   7   8   9   10

Notice that the first number is selected at random and that the second and third numbers follow a systematic-uniform pattern; the goal is to obtain the desired sample size from a fixed population in an efficient and unbiased manner.

As shown in the following example, systematic-random sampling provides a theoretically unbiased and efficient approach to estimating the expected value for the parameter number based on sampling a known fraction of the total population.

Goal: To make a theoretically unbiased estimate of the total number of objects in a population using systematic-random sampling.

Principle: Estimate the expected total object number, $E(N)$, as the product of the number in the sample and the reciprocal of the sampling fraction.

Approach: Make an estimate, $E(n)$, by counting the number of objects in one-third of the total, and multiplying the number sampled by the inverse of the sampling fraction.

$E(n) = n \times (3/1) = n \times 3$

Using this approach, there are three possible random starting points.

1. Starting random number = 1, then every third.

O   O   O   *O*   O   O   *O*   O   O   *O*
1   2   3   *4*   5   6   *7*   8   9   *10*

The number of objects sampled = 4 (objects 1, 4, 7,10). The sample estimate is

$E(n) = n \times 3/1 = 4 \times 3 = 12$

2. The starting random number = 2, then every third.

O   *O*   O   O   *O*   O   O   *O*   O   O
1   *2*   3   4   *5*   6   7   *8*   9   10

The number of objects sampled = 3 (objects 2, 5, 8). The sample estimate is

$E(n) = n \times 3/1 = 3 \times 3 = 9$

3. The starting random number = 3, then every third.

O	O	*O*	O	O	*O*	O	O	*O*	O
1	2	**3**	4	5	**6**	7	8	**9**	10

The number of objects sampled = 3 (objects 3, 6, 9). The sample estimate is

$E(n) = n \times 3/1 = 3 \times 3 = 9$

The results for estimates at three random starting numbers are shown below.

Starting random number	$n \times 3$
1	12
2	9
3	9
Sum	30

The mean value for the sample estimate, mean $E(n) = 30 / 3 = 10$. Recall that the expected (true) value for the parameter, $E(N) = 10$. Because $E(n) = E(N)$ =10, the sampling method is unbiased; the expected values for the sample estimate and the true value for the parameter are equal. This example illustrates a second important point about unbiased sampling: *With further sampling, the mean sample estimate, rather than any single sample estimate, will converge on the expected value.*

Sampling Biological Tissue

The previous discussions of random sampling illustrate the accuracy of independent (nonsystematic) and dependent (systematic) sampling. In biological tissue, both of these random sampling approaches lead to theoretically unbiased estimates at each level of the sampling hierarchy.

Because the goal in biological studies is to extrapolate findings from small samples to larger populations of unknown actual size, independent-random sampling is done at the individual or topmost level of the sampling hierarchy. At lower levels of sampling, that is, where the size of the reference space is fixed and entirely available for sampling, the possibility exists for systematic-random sampling that is both theoretically unbiased and efficient. The following example illustrates the use of systematic-random sampling in a fixed sample of biological tissue.

Goal: To obtain a systematic random sample of 5 sections from a total of 100 sections through a reference space.

Systematic-random approach: Divide the total number of sections by the fixed proportion, every kth, to generate the desired number of sections. For 5 sections from a total of 100, $k = 20$ ($100/20 = 5$). To ensure that the sample is random, use a random start to make the first section random in the interval 0 to 20, for example, 12. The first section in the series will be the twelfth section. After that section, take every twentieth section through the full series of 100 sections.

First section	= 12th
Second section	= 12 + 20 = 32nd
Third section	= 32 + 20 = 52nd
Fourth section	= 52 + 20 = 72nd
Fifth section	= 72 + 20 = 92nd

Thus, sections 12, 32, 52, 72, and 92 constitute a theoretically unbiased, systematic-random sample of sections through the reference space.

Efficiency of Systematic- versus Independent-Random Sampling

This section illustrates the relative efficiencies of systematic-random sampling compared with independent-random sampling. The example refers to a study to find the total volume of a reference space (cerebral cortex) in normal human brains. The sampling design can be summarized: Fixed human brains were cut in the coronal (transverse) plane into 18 slices (sections) of approximately 1 cm thickness each. The total set of slices contained the complete reference space. The total volume of the cerebral cortex, V_{CTX}, was found to be 632.5 cm^3 by sampling every section containing the reference space. The time required to estimate V_{CTX} on all 18 sections was approximately 4 hr.

Set	V_{CTX}(cm^3)	No. of sections	Time (hr)
1	632	18	4

We can take 632 cm^3 as the expected value for total V_{CTX}, $E(V_{CTX})$. Because the 4 hr to analyze all 18 sections is substantial, the question of efficiency arises. Would anything be lost in terms of accuracy by estimating the total volume using a smaller number of sections, in a fraction of the time?

Systematic-Random Sampling

To illustrate the efficiency of systematic-random sampling for tissue, a determination was made of total V_{CTX} for the human cerebral cortex. Instead of all 18 sections, every second section [sampling interval $(k) = 2$] was selected with a random start, resulting in two theoretically unbiased sample sets of 9 slices each through the reference space. Total V_{CTX} was estimated along with the standard deviation (SD), and the standard error of the mean (SEM $= SD/\sqrt{n}$) for the two estimates. The time to estimate V_{CTX} for each set was approximately 2 hr.

Set	$V_{CTX}(cm^3)$	No. of sections	Time (hr)
1	609	9	2
2	665	9	2

Mean V_{CTX} = 637 cm³
SD   = 40
SEM   = 28
Total time   = 2.0 hr/set × 2 sets = 4 hr

Analysis of two systematic-random samples of 9 sections each gives a mean sample estimate of 637 cm³. Systematic-random sampling required 4 hr to generate a mean sample estimate for V_{CTX} of 637 cm³, which closely approximates the value from counting all 18 sections (632 cm³). The small remaining difference ($637 - 632.5$ cm³) is accounted for by sampling error from point counting, as discussed in Chapter 10. Differences between the two individual estimates of V_{CTX} for sets 1 and 2 appear large (609 versus 665 cm³); however, this difference is less important when we consider that the expected range of normal values for this population is 550 to 725 cm³. Since both sample estimates fall within the expected range of the parameter, analyzing either set will provide an unbiased sample estimate of the parameter.

As shown by this example, rather than sampling 18 sections from each case, the more efficient approach for making a reliable estimate of a population parameter is to sample a larger number of individuals in the population in a systematic-random manner. In the next section we examine the effect of independent-random versus systematic-random sampling on the efficiency of sampling for the same set of sections.

Independent-Random Sampling

Six sets of 9 sections each were sampled in an independent-random manner from the total of 18 sections containing the entire reference space. The first set of 9 slices was taken and the total V_{CTX} estimated using an unbiased stereology method. This set was then replaced and the second set sampled and total V_{CTX} estimated, and so on, until 6 sets had been sampled and total V_{CTX} estimated for each set. The time required for estimation of V_{CTX} in each set was approximately 2 hr per brain.

Set	$V_{CTX}(cm^3)$	No. of sections	Time (hr)
1	723	9	2
2	779	9	2
3	666	9	2
4	625	9	2
5	593	9	2
6	702	9	2
		$\Sigma = 54$	12

Mean V_{CTX} = 681 cm^3
SD   = 68
SEM   = 28
Total time = 2 hr/set $\times$ 6 sets = 12 hr

Compare this estimate of a mean of 681 cm^3 from independent-random sampling, which required three times the time (4 versus 12 hr) and effort (54 versus 18 slices) to estimate mean V_{CTX} as 637 cm^3 as shown previously for systematic-random sampling. As shown in Figure 3-1, systematic-random sampling more quickly converges on the true value and is therefore more efficient than independent-random (pure random) sampling.

It is important to emphasize that both systematic and independent forms of random sampling are accurate; with further sampling, both will generate mean estimates that converge on the expected value for the parameter. The critical difference is efficiency, as expressed in Table 3-8 in terms of precision units (SD) per unit of time.

In the example given here, systematic-random sampling is 66% more efficient than independent-random sampling at capturing the within-sample variation in total V_{CTX}. Fewer sections are analyzed in less time and capture a greater amount of variation in the population. Thus we can understand why systematic-random sampling is more efficient at capturing the vari-

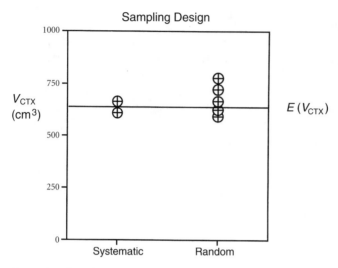

Figure 3-1. Sampling designs showing expected value for total cortex volume in cubic centimeters using systematic-random or pure random sampling

ation in biological structures than independent-random sampling. Systematic-random sampling collects data along a single axis in a sequential manner, beginning with the first section in the series and continuing through the entire structure without sampling the same region more than once. Thus it is often referred to as dependent sampling to illustrate the dependence of the magnitude of the parameter on the previous region and the next region within the reference space.

In contrast, independent-random sampling does not sample along a single axis; instead, data points are independent of the previous or subsequent data points. The conclusion is that although both strategies will eventually arrive at a theoretically unbiased estimate of the parameter, the systematic-random approach arrives at this point with less sampling, time, effort, and cost.

Table 3–8 Comparison of Efficiencies for Systematic-Random and Independent-Random Sampling

Sampling design	No. of slices	SD	Time (hr)	Efficiency (SD/hr)
Systematic-random	18	40	4	10.0
Independent-random	54	68	12	5.7

Summary

Efficient and theoretically unbiased sampling approaches are a relatively recent development in the history of stereology. When mathematicians and statisticians joined the ISS in the 1970s, their first contribution was to show that biological tissue could be efficiently sampled without bias. The two essential parts to this approach are to make an unambiguous definition of a biologically interesting tissue or population of objects and then sample these structures in a systematic-random manner. Today, systematic-random sampling is the essential first step in an efficient and theoretically unbiased stereology design.

4

Geometric Probes

This chapter discusses the use of geometric probes for theoretically unbiased estimates of first- and second-order stereological parameters. By the 1970s the weakness inherent in traditional model-based approaches had become evident, leading mathematicians and biologists to devise theoretically unbiased techniques using geometric probes to sample and estimate the true dimensions of biological objects on tissue sections. The design of these probes ensured that intersections between the probe and the features of interest in the tissue would be proportional to the parameter of interest, without futher assumptions about the size, shape, or orientation of the biological objects. Today, theoretically unbiased geometric probes are used in combination with systematic-random sampling to make accurate, precise, and efficient estimates of biological structures.

Theoretically unbiased stereological estimators have been developed for the estimation of volume (size), number, surface area, and length of biological objects from their appearance on the cut surface of tissue sections. These estimators use geometric probes that avoid assumptions and models that can introduce bias into sample estimates. Modern stereological estimators are based on principles of stochastic geometry and probability theory. In subsequent chapters we introduce a number of theoretically unbiased estimators for each of the first-order stereological parameters of interest. Estimators are rules that are used in conjunction with a theoretically unbiased geometric probe, for example, the 3-D disector, to estimate a particular parameter. Because geometric probes and the parameter they are designed to estimate vary in their dimensions, it is important to select the correct probe–parameter combination for a particular study.

New Stereology and Geometric Probes

The success of new stereology in biological applications lies in its strong theoretical foundation in stochastic geometry and probability theory. The goal of stochastic geometry is not to "fit" the morphological feature of interest into a particular model, but rather to "capture" variability of the parameter in a single, stable estimate. The probability theory underlying new stereology is similar to that in another area of applied mathematics—inferential statistics. In both cases, estimates from a small number of individuals sampled at random from the target population are used to make inferences about population parameters. For inferential statistics, we are interested in the probability of statistically significant differences between groups. If all known sources of bias are avoided during the estimation of sample parameters in a small sample, the results can be extrapolated to population parameters. For both inferential statistics and new stereology, the probability of generating statistically meaningful conclusions is increased when all known forms of systematic error are avoided.

Geometric probes intersect objects in a tissue with a probability that is directly related to the magnitude of the parameter of interest. To be theoretically unbiased, the number of dimensions in the parameter of interest and the probe must sum to 3 (i.e., $probe_{dim} + parameter_{dim} \geq 3$). For instance, to estimate total number, a 0-D parameter, we use a 3-D probe called a disector; for total surface, a 2-D parameter, we use a 1-D line probe; for total length, a 1-D parameter, we use a 2-D planar probe; and for total volume, a 3-D parameter, we use a point grid, 0-D probe.

In this chapter we review the features of stereological probes used to make theoretically unbiased estimates of biological structures. These geometric probes were specifically developed with one idea in mind: to avoid assumption- and model-based sources of bias when estimating stereological parameters of objects in 3-D. To achieve this goal, the probe should intersect the objects in the tissue with a probability that is directly related to the expected value of the parameter of interest. When they are combined with systematic-random sampling as discussed in Chapter 3, theoretically unbiased geometric probes allow stereological parameters to be estimated without bias and with efficiency. In Chapter 10 we review methods to optimize this efficiency based on the variation observed in a particular parameter following an initial pilot study.

The Needle Problem

The needle problem, posed in 1777 to the Royal Academy of Sciences by Count Georges-Louis Leclerc de Buffon, stands as a significant landmark in

Figure 4-1. Buffon's needle problem

modern stereology (Figure 4-1). He presented the problem as follows: *What is the chance a needle tossed in the air will intersect lines on a parquet floor?* Restating Buffon's needle problem in terms of modern stereology: *What is the probability that a randomly placed line will intersect a grid of parallel lines?*

In the tradition of eighteenth-century science, Buffon provided the answer: The probability of needle–line intersections is directly proportional to the ratio of the needle length and the distance between the lines. For either a longer needle or a smaller distance between lines on the floor, the probability of an intersection increases. In the 1960s, the needle problem became of interest for applications of new stereology to biological tissue with the realization that if one quantity is known, for example, the distance between lines on a line probe, the length of the needle can be estimated in a theoretically unbiased manner from the probability of an intersection. That is, by counting the number of intersections from random placement of the needle in proximity to the line probe, the expected value for the length of the needle can be estimated according to the formula in Figure 4-2, where $p(I)$ is the probability of an intersection, the ratio of intersections to tosses; L is the line length; and d is the distance between lines on floor.

The constant, $2/\pi$, randomizes all possible intersection angles between the needle and the lines on the floor. All possible angles between the needle and the line are randomized by tossing the needle in the air and allowing it to hit the line probe on the floor at random; the factor $2/\pi$ converts the rela-

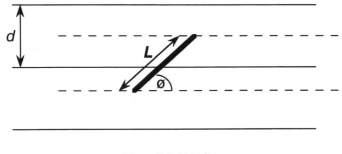

$$p\,(I) = (2/\pi)(L/d)$$

If $p(I) = I/N$, then

$$L = (\pi/2)(I/N)\,d$$

Figure 4-2. Mathematical representation of the needle problem. (From Weibel, 1979.)

tionship from a proportionality to a quantitative value. In Chapters 6, 7, and 8 we return to Buffon's needle problem to discuss its applications for estimating the length and surface area of biologically interesting objects.

The theoretical use of the probability of intersections between a needle tossed at random onto a line probe stands as one of the first applications of probability theory and stochastic geometry to the estimation of a first-order stereological parameter. It would be another two hundred years before this idea would be applied to the estimation of stereological parameters of biological importance.

Biological Applications of Probability Theory

In the late 1960s two biological research groups had the opportunity to use the theoretical approach of geometric probes and probability theory, as exemplified by the needle problem, to estimate the total surface area of endoplasmic reticulum (ER) in the rat liver. Both studies used the same stereological technique to estimate the same parameter (surface area) in the same reference space (rat liver): a 1-D probe (line grid) to estimate a 2-D parameter (surface area). In both cases, random intersections occurred between the probe (line grid) and the biological feature of interest (the surface of the ER). The only difference between these studies was practical: magnification. In both cases, the prediction was that the number of probe-to-surface intersections would be proportional to the total surface area of ER in the reference space, with no further assumptions. Surprisingly, the results of the two stud-

Table 4–1 Comparison of Two Results for Total
Surface Area of Endoplasmic Reticulum in Rat
Liver Cells

Magnification	Value	Source
12,000	6 m^2/cm^3	Loud (1968)
80,000	11 m^2/cm^3	Weibel et al. (1969)

ies did not agree; in fact, they were not even close, as shown by the values in Table 4-1.

Both groups checked and rechecked their data, looking for an explanation for the discrepancy. However, they could not find one. To biologists not predisposed to new stereology, including those annoyed by its claims of unbiasedness and assumption-free methods, these results were interpreted as evidence that stereology was not immune to the types of discrepancies that regularly occur with assumption-based methods.

One of the researchers involved in the investigations, Professor Ewald Weibel in Bern, Switzerland, decided to follow up on these studies. As shown in Figure 4-3, Professor Weibel's work demonstrated that progressively higher resolution was positively related to higher sample estimates for the total surface area of ER in rat liver. As in the earlier work by Loud (1968), Weibel and his collegues (1969) could find no explanation for this result. Later in this chapter we will revisit this caveat, known as scale dependance, which illustrates an important limitation of the theoretical unbiasedness of new stereology. In the next section we use the ideas developed so far to describe a stereological approach for estimating the total area (ΣA) of the cut surface of sections.

The Delesse Principle

The French mining engineer Auguste Delesse was interested in quantifying particular phases in geologic samples. To visualize these phases, he used a heavy saw to cut random sections through the phases in rock specimens. Delesse sought to prove what today seems simple: that a quantitative relationship exists between the phase volume in the entire sample and the phase area on the cut surfaces. An analogous problem in biology is the relationship between the volume of objects on a random section through tissue and the volume of the objects within the tissue.

To test the hypothesis that A_{ref} is directly related to V_{ref}, Delesse compared the area of the phase on the cut surface with the volume of the phase

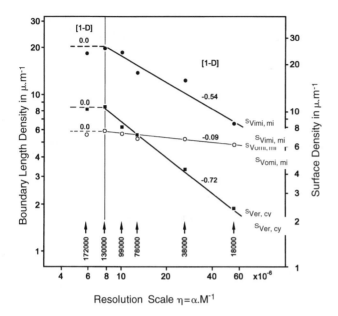

Figure 4-3. Weibel's data for scale dependence showing the relationship between surface density (boundary length) and resolution: A greater surface area is expected at higher magnifications. (From Paumgartner et al., 1981.)

through the entire rock. He used transparent drawing paper to trace a particular phase on the rock's cut surface, then cut out the outlined areas and weighed them. From these measurements he was able to determine A_{phase} within a defined reference area of rock, A_{ref}; he expressed this ratio as the area fraction, A_{phase} / A_{ref}. These data also permitted him to determine the number of deposits of a particular phase, N_{phase}, as a fraction of reference area, A_{ref}; he expressed this quantity as the areal density, N_{phase} / A_{ref}. These terms quantified the phase on the 2-D cut surface of the rock. To relate these quantities to the corresponding 3-D volumes, Delesse carried out a careful reconstruction of the rock from serial thin sections. These studies produced values for the volume fraction of the phase through the rock, V_{phase} / V_{ref}, and the number of deposits per unit of volume or the volume density, N_{phase} / V_{ref}.

Delesse compared the 2-D quantities of particular phases on the cut surface with the corresponding 3-D quantities in the rock. His analysis confirmed that the area fraction was indeed proportional to the volume fraction, $A_{phase} / A_{ref} = V_{phase} / V_{ref}$. Furthermore, Delesse found that the areal density of phase deposits was not related to their volume density, $N_{phase} / A_{ref} \neq N_{phase} / V_{ref}$. These findings today comprise the famous Delesse principle, a

classical demonstration of the relationship between 2-D quantities and corresponding parameters in 3-D. The formula for the areal fraction is stated as
follows:

area fraction = volume fraction

$$A_{obj} / A_{ref} = V_{obj} / V_{ref}$$

where:

A_{obj} = area of the object's profile (2-D)
A_{ref} = area of the reference space (2-D)
V_{obj} = volume of the object (3-D)
V_{ref} = volume of the reference space (3-D)

The relationship between area density and volume density is

areal density ≠ volume density

$$N_{pro} / A_{ref} \neq N_{obj} / V_{ref}$$

where N_{pro} is the number of object profiles (2-D), N_{obj} is the number of objects (3-D), and V_{ref} is the volume of the reference space (3-D).

The Delesse principle established the basis for several important concepts underlying the stereological analysis of biological objects on tissue sections (Figure 4-4). It relates the area of an object on 2-D planes to its corresponding volume in 3-D. Furthermore, the Delesse principle shows that
single planes (sections) do not provide an effective probe for number, since
$N_{phase} / A_{ref} \neq N_{phase} / V_{ref}$. As discussed in the next section and in greater

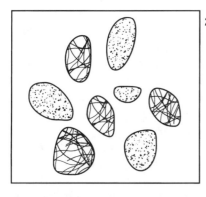

23% of field contains structure A but. . .

$$N_A \neq N_V$$

34% of area is structure A.

$$A_A = V_V$$

 Structure A   Structure B

Figure 4-4. Delesse principle demonstrated that for random sections through objects
containing particles of interest, the number of particles per unit of area is not equivalent
to the number of particles per unit of volume ($N_A \neq N_V$); however, the area of particles
per unit of area is equivalent to the volume of particles per unit of volume ($A_A = V_V$).

detail in later chapters, the Delesse principle is central to the analysis of biological structure on 2-D sections. Unfortunately, biologists would not become aware of this work until the members of the geological and biological
disciplines began to attend the same meetings on stereology after the formation of the ISS in 1963.

Grid Constants

Recall from the needle problem that the probability of an intersection between a probe and the structure of interest is related to one unknown quantity, the amount of structure, and one known quantity. In the needle problem, if we know either the length of the needle or the distance between the
lines on the floor, we can estimate the unknown quantity in a theoretically
unbiased manner using the laws of probability theory. In practice, modern
stereology uses exactly the same approach to estimate the unknown quantity
of a structure in a defined reference space. The known quantity in our equation is called a grid constant. The grid constant adjusts the probability formulas for the amount of the probe used to analyze the reference space. The
greater the amount of the probe, the greater the probability of an intersection. Because the only unknown in the equation must be the amount of structure in the tissue, the grid constant is taken into account in the formula, as
shown in the following example.

> *Goal:* To observe the inverse relationship between the sum of inter-
> sections (ΣI) and the grid constant to estimate a given amount of struc-
> ture, $E(x)$.
>
> *Formula:* $E(x) = \Sigma I \times$ grid constant
>
> grid constant = number probes per unit of area

$E(x)$	ΣI	No. of probes	Area (μm^2)	Grid constant (probe/μm^2)
100	10	100	10	10
100	100	1.0	1.0	1.0
100	0.1	1000	1.0	1000
100	150	100	150	0.67

The table illustrates the direct relationship between the amount of the
probe per unit of area and the number of probe–parameter intersections
counted. The grid constant is the areal density of a particular grid of geometric probes, the ratio of the amount of probe to the area on the grid.

The Coastline of Britain

Recall the discrepant studies by Loud and by Weibel et al. on the total surface area of ER in the rat liver. These studies both used theoretically unbiased stereology, but failed to arrive at similar results as predicted by stochastic geometry and probability theory. Follow-up studies by Weibel showed that these estimates were scale dependent, that is, the expected value increased in direct proportion to the level of resolution.

In 1977, at the two hundredth anniversary of the presentation of Buffon's needle problem, an international meeting of stereologists was held in Paris to discuss the impact of probability theory on stereology. One speaker at this meeting was Benoit Mandelbrot (1924–). It was Mandelbrot's presentation that provided the remarkable insight necessary to understand the discrepancies experienced by Weibel and Loud. Specifically, Mandelbrot showed why surface area estimates in a defined reference space are in fact greater at higher resolution. The reason is that the surfaces and boundary lengths of biological objects behave like *fractals* (Figure 4-5). According to Mandelbrot (1983): *Nature exhibits not simply a higher degree but an altogether different level of complexity. The number of distinct scales of length of natural patterns is for all purposes infinite.* That is, the closer one looks, the more biological surface is present.

Mandelbrot's (1967) fractal analogy, "the infinite coastline of Britain," provided an explanation for Weibel's empirical observations of the scale dependence between surface area and resolution. Imagine tracing the coastline of Great Britain while orbiting the Earth in the Space Shuttle. A problem arises as we move closer and closer to the coastline. If we quantify the length

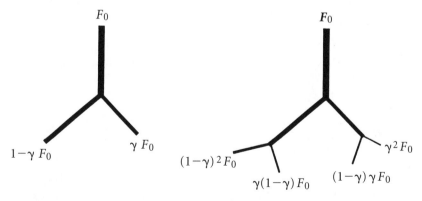

Figure 4-5. Mandelbrot's fractal provided a mathematical basis for Weibel's observations of the scale dependence of boundary length and surface area. (From Glenny and Robertson, 1991.)

of the coastline of Great Britain while we are, say, 100 miles from Earth, and then repeat this exercise at a closer distance, say, 30,000 feet, after correcting for distance, the values we would calculate for the actual length of the coastline would differ. The area estimate made closer to the Earth will be higher than the one made at a greater distance. As we move progressively closer to the coastline of Great Britain, the apparent length of its boundary will continue to increase.

According to Mandelbrot, the boundary length becomes an elusive notion that slips between the fingers as one attempts to grasp it. Like the coastline of Britain, the surface area and boundary length of the ER in the rat liver vary along a fractal dimension. It is important to know this critical feature of surface area and length when comparing estimates made at different magnifications, as in the case of the studies by Weibel et al. and Loud in the 1960s. In these cases, we must be careful to compare surface area and length estimates at similar magnifications (resolution); otherwise we are likely to misinterpret differences arising from the scale dependence of fractal dimensions as biological differences in length and surface area.

Of the four first-order stereological parameters, surface area and length vary along fractal dimensions—the closer you look, the more there is. The parameters of number and volume do not vary along fractal dimensions; thus we can compare these parameters at different magnifications. For example, two or more studies of the same reference space made to estimate the total cell number using good stereology should obtain similar results, as shown in Figure 4-6. Though they were carried out by geographically separate groups without contact or knowledge of each others' work, these studies obtained remarkably similar results, as expected, regardless of magnification.

The above examples are not meant to imply that theoretically unbiased methods always produce findings that are consistent across research groups. Indeed, estimates from different research groups may be based on slightly different populations, which may or may not show coincident results. Unlike the fractal dimensions of surface area and length, however, differences observed in the number and volume of objects are scale independent. When parameters of number and volume are quantified using unbiased methods for the same population and reference space, the results are not expected to vary because of scale.

Summary

Modern stereological approaches use unbiased techniques based on stochastic geometry and probability theory. The application of stochastic geometry

Total number of neuromelanin-containing neurons in the human nucleus coeruleus

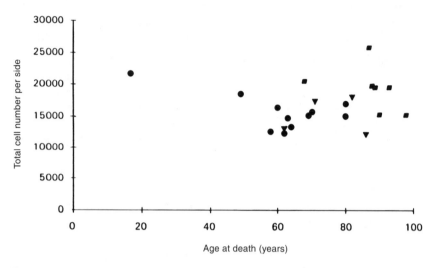

Figure 4-6. Results from independent studies using unbiased stereology on the same cell population in autopsied human brains. These studies confirmed no significant cell loss during normal aging. (Data from Mouton et al., 1994, and Ohm et al., 1997.)

to nonclassically shaped biological structures requires that the sum of dimensions in the probe–parameter combination equal at least 3, the number of possible dimensions of biological tissue. The use of probability theory requires that random intersections occur between the probe and the biological structures of interest. When these two requirements are met, geometric probes used in conjunction with systematic-random sampling permit theoretically unbiased estimates of all first-order stereological parameters of biological objects. In contrast to number and volume, the parameters of length and surface area are scale dependent; sample estimates are comparable only at the same resolution.

5

Bias in Estimating Number

The avoidance of stereological bias in estimating total object number in a defined reference space is discussed in this chapter. Stereological bias is created when a 2-D sampling plane (knife) cuts 3-D objects into thin histological sections. The goal of stereology is to overcome this and other sources of bias that can introduce systematic error into sample estimates. Recognition and avoidance of these sources of error requires a thorough understanding of the bias introduced by assumptions, models, and correction factors.

To **this point** we have discussed the importance of avoiding stereological bias in estimating first-order stereological parameters. In this chapter and the next, we focus on a specific parameter, number, which is perhaps the most common parameter of interest to bioscientists. This chapter reviews the assumption- and model-based sources of stereological bias related to observing 3-D objects (e.g., cells) as 2-D profiles on tissue sections. The tissue-processing artifacts that can introduce nonstereological bias into estimates of total number are discussed in Chapter 9.

N_A versus N_V

In Chapter 4 we introduced the Delesse principle, which was first proposed in 1847, as a means of understanding the differences between the number of profiles per unit of area (N_A) and the number of objects per unit of volume (N_V). Delesse showed that $N_A \neq N_V$, for a random section through a reference space containing 3-D objects of interest.

Further examination of this problem reveals the underlying basis for this inequality. To understand the differences between N_A and N_V, we should consider the distribution underlying the desired parameter, N_V, in a defined population of biological organisms. By sampling individuals from this distribution and estimating N_V, then repeating this process for several more individuals sampled at random from the population, we are attempting to estimate the central tendency and variation of the distribution. The number of individuals that must be analyzed to estimate the true central tendency and variability of this distribution depends on how much variation exists in these parameters at the population level. Based on experience with other populations of biological objects under similar selective pressures, we expect a stable estimate of the total distribution after a minimum of five individuals, with about ten individuals being adequate in most cases for reliable estimates.

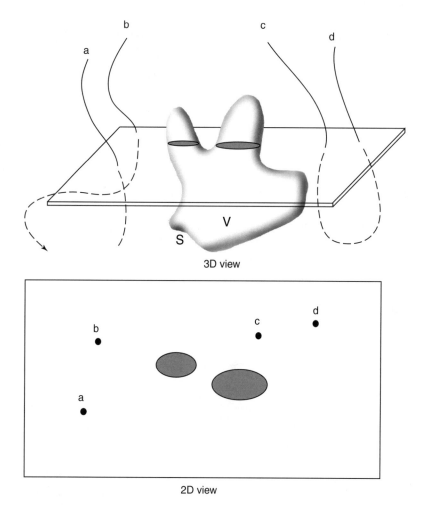

Figure 5-1. Three-dimensional objects appear distorted in 2-D when sampled with 2-D planes. S, surface area; V, volume.

Table 5–1 3-D and 2-D Parameters

Parameter	3-D	2-D
Volume	Volume (3-D)	Area (2-D)
Area	Area (2-D)	Boundary (1-D)
Length	Length (1-D)	Point (0-D)
Number	Number (0-D)	—

Now consider the distribution for the number of object profiles per unit of area, N_A. We can only estimate this distribution using assumptions about the size, shape, and orientation of the objects in the tissue. The presence of stereological bias changes the number distribution in sample estimates by an unknown magnitude and direction from the true number distribution. As shown in Table 5-1 and Figure 5-1, tissue sectioning and low optical resolution create a mismatch between the apparent dimensions of stereological parameters in 3-D and 2-D. A 2-D sampling knife (a microtome) passing through tissue containing 3-D objects selects arbitrary planes through the object. On tissue sections, 3-D objects appear as profiles in a 2-D area; the number of 3-D objects cannot be seen in a 2-D section (Figure 5-2).

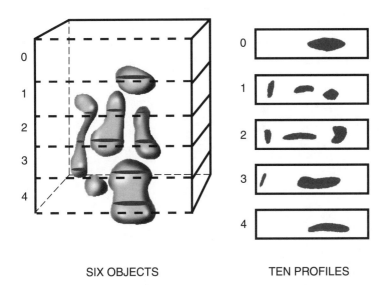

SIX OBJECTS   TEN PROFILES

Figure 5-2. The corpuscle problem. Six objects with arbitrary size, shape, and orientation appear as ten profiles on 2-D sections.

The Corpuscle Problem

In the mid-1920s a Swedish mathematician named S. D. Wicksell studied the
number of objects on a tissue section and concluded, like Delesse in his studies of phases in geologic specimens, that in tissues also, $N_A \neq N_V$. Wicksell
called this inequality the corpuscle problem. Through careful 3-D reconstruction and statistical modeling with spheres and ellipsoids, he identified
three features—size, shape, and orientation—as the critical determinants of
whether a cell in a defined region of tissue will be sampled by a knife blade
passing through the tissue.

Wicksell showed that larger cells, cells with more complex shapes, and
cells with their long axis perpendicular to the sectioning plane, have a higher
probability of being hit by a 2-D sampling plane, and therefore of appearing
as a profile on the tissue section and being counted. By modeling cells as
spheres and ellipses, Wicksell proposed an early model- and assumption-
based correction factor in an attempt to overcome the corpuscle problem.

$$N_V = N_A / D$$

where N_V is the number of objects per unit of volume (μm^{-3}), N_A is the
number of objects per unit of area (μm^{-2}), and D is the mean diameter of
the object (μm).

Wicksell's correction formula, like several subsequent attempts to overcome the corpuscle problem, required certain assumptions and models derived from classical geometry (Table 5-2). By the advent of modern stereology in the early 1960s, the literature from the biological and materials
sciences was replete with new correction factors, corrections of previous cor-

Table 5– 2 Correction Factors for Number of Object Profiles

Researchers and date	Correction factors
Wicksell, 1925	$N_V = N_A/2r$
Agduhr, 1941	$N_V = N_A[2m \times T - 2r)/T_m(m + 1)]$
Floderus, 1944	$N_V = N_A[T/(T + 2r - 2k)]$
Abercrombie, 1946	$N_V = N_A[T/(T + 2r)]$
Weibel and Gomez, 1962	$N_V = K(N_A)^{3/2}/\beta(V_V)^{1/2}$
Ebbeson and Tang, 1965	$N_V = (N_{A1} - N_{A2})/T_1 - T_2$

Note: N_V = actual object number per unit of volume; N_A = number of object profiles counted in
area A; T = section thickness; $2r$ = diameter of structure; $2k$ = smallest detectable part of the
structure; K = size distribution coefficient; β = shape coefficient; m = number of sections in
which structure wholly or partly falls.

rection factors, and numerous modifications based on new models and assumptions.

The problem with correction factors is not that the equations are incorrect. On the contrary, if the assumptions of the formulas are met, the equations are correct. However, the models required to make the assumptions true do not occur in biological objects. As discussed in Chapter 1, a problem exists at the core of these formulas: Biological objects do not fit classically shaped models.

We can also understand the inequality $N_A \neq N_V$ using a more intuitive approach. Recall that the number of potential dimensions within tissue that can be occupied by cells is 3. Because all three of these dimensions could contain cells, to quantify the number of cells in tissue, all three dimensions must be probed. Therefore, counting the number of cells in tissue requires a 3-D probe. A probe with fewer than three dimensions will fail to sample all of the possible spaces within a tissue that could be occupied by cells. When we use a tissue section to sample profiles for cell number, we are in effect attempting to count the number of cells using only a 2-D probe. As illustrated in the following section, to overcome this problem, one of the most widely used correction factors, the Abercrombie correction, requires a number of assumptions that do not apply to biological objects.

The Abercrombie Correction

The Abercrombie formula is as follows:

$$N_v = N_A [T / (T + D)]$$

where:

N_V = corrected estimate of object number per unit of volume (numerical density)
N_A = uncorrected profile estimate per unit of area
T = mean thickness of the section (μm)
D = mean diameter of the objects measured perpendicular to the direction of sectioning (μm)

Use of the Abercrombie correction formula is simple and straightforward: (1) Count the number of profiles on a tissue section (n); (2) estimate the mean section thickness (T) and the mean object diameter (D); and (3) convert the profile estimate (N_A) to the corrected estimate of object number (N_V).

Like Wicksell's correction, the Abercrombie correction sought to overcome the corpuscle problem. However, the bias created by the corpuscle problem remains; a 2-D sampling plane is being used to count the number of 3-D objects. In addition to not overcoming this bias, the Abercrombie formula introduces further bias with the assumption that the observer has a reliable estimate of mean *D*, the mean diameter of objects in the reference space of interest. However, mean *D* cannot be determined from routine tissue sections, at least not by cutting along a single axis, as is done during the preparation of tissue for morphometry. Making a reliable estimate of mean *D* requires a far more rigorous investigation, thus losing the major attraction of the correction factor—convenience. In practice, the Abercrombie correction is used with a rough estimate of mean *D*, which is usually found by measuring the long axis of an arbitrary number of object profiles on the section. Obviously this estimate will be heavily biased by the profiles selected. Because larger objects are more likely to be cut with a 2-D sampling plane, the profiles will reflect a biased sample in the reference space. Furthermore, for nonspherical objects, the mean *D* will change according to the direction of sampling. Thus, to estimate *D* as required for the Abercrombie formula, measurement of profiles on routine histological sections must assume that the objects of interest are spheres. If one is counting objects with a reasonably spherical shape, for example, nucleoli, this assumption is met reasonably well; *D* can be estimated for use in the formula. However, as the objects of interest depart from pure sphericity, bias introduced by the correction increases.

A second source of bias in the Abercrombie correction formula arises from the mean section thickness, *T*, which is estimated at the time the tissue is sectioned, that is, from the microtome setting. Microtomes are more or less designed to deliver sections with a high level of precision. As shown in the following examples, a difference in mean section thickness of between 40 and 20 μm will introduce a significant difference in the corrected value of N_V.

Example 1: Assumed instrument setting for mean section thickness = 40 μm. *D* = 3 μm; *n* = 100 profiles counted per unit of area.

$N_V = 100[T / (T + D)] = 100 \times [40/(40 + 3)] = 100 \times (40/43) = 93$ cells

Example 2: Actual mean section thickness = 20 μm. *D* = 3 μm; *n* = 100 profiles counted per unit of area.

$N_V = 100[T / (T + D)] = 100 \times [20/(20 + 3)] = 100 \times (20/23) = 87$ cells

The difference related to the assumption of section thickness = (92 − 87/92) × 100 = 5.4%.

In addition to the biases associated with the assumptions in mean *D* and mean *T*, and the assumption that the cells are spheres, the Abercrombie correction assumes that objects are oversampled by a 2-D sampling plane; the corrected estimate of N_V is less than the number of profiles counted per unit of area. Objects have a reduced probability of being sampled if they are smaller, less complex, and do not have their long axis perpendicular to the direction of sampling. These factors result from the variability of objects in tissue; we have no way to quantify their number. In theory, bias from this source should be added to corrected values of N_V.

Finally, between-group differences in size, shape, and orientation of cells may be significant if the same correction factors are used to correct data from groups subjected to different treatment or with different characteristics (e.g., aging versus young, differences in drug dosage or duration of treatment, time to recovery). If these effects systematically change the size of objects in one group, bias caused by under- and oversampling will differ between the groups in an unpredictable manner.

Probabilities of bias are multiplicative rather than additive; that is, the net probability of bias from all the assumptions is the *product* of the probabilities of individual biases. There is wide variability in the accuracy of assumptions and models for the diverse tissues where the Abercrombie and other correction factors are used. One rarely encounters populations of biological objects that satisfy the models and assumptions of the Abercrombie correction formula. Thus, given that multiple assumptions show a high probability of introducing bias, correction factors also have a high probability of introducing stereological bias into the estimate and ultimately the literature. Rather than correcting data, correction formulas simply increase the probability of introducing stereological bias into morphological data.

In addition to bias from the corpuscle problem, another bias commonly encountered in estimating number occurs when densities are reported in place of total number in a defined reference space. Unlike a first-order stereological parameter, density is more difficult to understand and can lead to a variety of misinterpretations. Because of the wide use of density-based estimators of object numbers before the advent of new stereology, this approach is considered in greater detail in the next section.

Why *Not* Density?

Numerical density is the number of objects per unit of reference space (area or volume), an assumption-based estimator. Density is a ratio estimator because it reflects the ratio of a numerator and denominator (see Table 5-3). Like all ratios, a density estimate is influenced by variation in both parameters. For this reason, density estimates are misleading estimates of the number of objects in a defined reference space. For instance, the nation of Monaco

Table 5–3 Density Estimators for Number

Ratio estimator	Units	Numerator	Denominator (units)
Area density (N_A)	(μm^{-2}, mm^{-2}, etc.)	Object	Per unit of area (μm^2, mm^2, etc.)
Volume density (N_V)	(μm^{-3}, mm^{-3}, etc.)	Object	Per unit of volume (μm^{-3}, mm^{-3}, etc.)

along the southern border of France has one of the world's highest population densities because the entire population of 30,000 persons lives in an area of only 0.75 square miles (1.9 km^2)!

The units of density estimators are less straightforward than absolute number. Total number is a 0-D parameter and has no units. As shown in Table 5-3, units of densities can be expressed as inverse area (μm^{-2}, mm^{-2}, etc.) and inverse volume (μm^{-3}, mm^{-3}, etc). Density estimates may report a 30% change in the inverse volume units, which is clearly not the same as a 30% reduction in total number. Densities convey ambiguous information about changes in first-order stereological parameters. Estimates of density may change as a result of changes in the number of objects or changes in the reference space, or both. Interpretations of density estimators require assumptions about changes in the reference space. As shown in Figure 5-3, changes in the reference space can change the numerical density of objects in a defined reference space, in the absence of a change in the total number of objects.

The Reference Trap

The first-order stereological parameters relate to absolute, rather than relative, parameters. That is, these parameters are not expressed in relative terms, such as number per unit of area, length per unit of volume, and so on. Estimating absolute values, in contrast to relative values such as density,

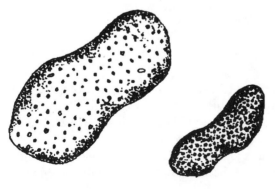

Figure 5-3. Schematic example of the reference trap for ratio estimators. Changes in the reference volume in the absence of a change in object number lead to biased estimates of density (number of objects per unit of volume).

avoids introducing bias when changes occur in one part of the parameter (e.g., volume of reference space) that are independent of changes in the other part of the parameter. For example, changes in N_V may result from changes in the number of objects or in the volume of the reference space in which these objects occur; both changes will be reflected in the density estimate. Such changes can be caused by a number of factors. For instance, when biological tissue is processed for stereological analysis, it is subjected to reagents that can change the volume of the reference space (shrinkage or expansion) without changing the number of objects. Biological tissues also undergo changes after separation from their blood supply, causing cellular autolysis and putrefaction, which in turn can cause uneven changes in the parameters of density estimates. These artifacts can be dramatically minimized by tissue processing, beginning with immediate fixation with aldehydes and alcohols. Fixation can cause water and ions to move out of tissue, however, leading to nonrandom changes in stereological parameters.

Thus, when we attempt to quantify a first-order parameter in relative terms, because of nonstereological sources of bias inherent in tissue processing, we risk falling into the reference trap. The reference trap applies not only to numerical density (number per unit of area or volume) but also to length density, surface density, and volume density. An estimate of density is valid only if it is assumed that only the parameter of interest is changing and that the reference space remains constant, an unlikely assumption for biological tissue processed for stereological analysis. Because of the reference trap, estimates of biological parameters expressed as density and other ratio estimators are considered to be assumption based; that is, they are likely to be accurate only when certain unlikely assumptions are met. Unless one takes the time and effort to accurately verify assumptions about changes in the reference space for all samples in an estimate, the use of ratio estimators such as density can easily result in biased estimates and incorrect conclusions.

The disector method provides a theoretically unbiased estimate of numerical density, N_V. However, N_V is a ratio, the number of objects per unit of volume, whether estimated by a theoretically unbiased method or not. Like all ratios, changes in N_V may reflect changes in the numerator or the denominator, or both. Relying on N_V as an estimate of number is analogous to using miles per hour as an estimate of distance. Does a car traveling 75 miles per hour travel farther than a car traveling 50 miles per hour? The answer is yes, assuming the slower car doesn't travel for 10 hr compared with 3 hr for the faster car.

There are certain cases in which the reference trap is not a problem. For instance, recall Delesse's studies of rocks, which showed that area fraction equals the volume fraction ($A_A = V_V$). If the area fraction of a phase in specimen A is twice that for specimen B, the volume of the phase in specimen A is twice that for specimen B, with no further assumptions. Despite being based on ratios, this conclusion is unaffected by the reference trap because, unlike processing of biological tissue, processing of rock specimens does not cause changes in either of the parameters in the ratio. With the exception of bone, exoskeletons, and other hard tissues, physiocochemical reactions, including agonal (death-induced) changes and processing of tissue, can cause significant alternations in either parameter in a biological ratio.

Therefore, ratios of biological parameters are only valid under broad, and for the most part false or unverifiable, assumptions that different tissues undergo similar artifactual changes. Ignoring this caveat, and falling for the reference trap can lead to misinterpretations and inaccurate conclusions when analyzing morphological parameters of biological tissue.

Summary

In the past four decades modern stereology has identified the essential factors that introduce stereological bias into estimates of number on tissue sections. Delesse quantified the bias inherent in 2-D sampling planes for counting numbers of 3-D objects ($N_A \neq N_V$). Three-quarters of a century later, Wicksell identified the corpuscle problem arising from variation in morphology (size, shape) and orientation for biological objects in tissue. Throughout the twentieth century, a variety of questionable assumption- and model-based correction formulas were developed to overcome the inherent bias created by sampling 3-D objects with a 2-D sampling plane. Other sources of stereological bias for total number include reporting changes in terms of density (the reference trap) rather than absolute parameters.

6

The Disector Principle

The disector principle is the first theoretically unbiased method used to estimate total number of objects per unit of volume (numerical density, N_V) on tissue sections. The disector is a 3-D geometric probe for counting numbers of objects (cells) with a probability that is unaffected by the size, shape, or orientation of the objects. In combination with precautions to avoid edge effects, the disector method permits total numbers of cells to be estimated without assumptions, models, or correction factors. Practical applications include counting objects with two physical planes (physical disector), two optical planes (optical disector), and optical planes in conjunction with the fractionator sampling scheme (optical fractionator).

C*hapter 5* outlined the major sources of stereological bias in estimates of total number. Five steps are necessary to avoid these pitfalls:

1. Randomly select individuals from the population.
2. Identify the border of a defined reference space.
3. Perform systematic-random sampling of objects in the reference space.
4. Count the total number of objects in a known volume of reference space.
5. Scale the local estimate to the complete reference space.

This chapter focuses on step 4, how to count the total number of objects in a defined reference space. There are three prerequisites for estimation of the total number of 3-D objects in a theoretically unbiased manner:

1. The sum of probe(dim) + parameter(dim) must equal at least 3.
2. Every object must have an equal probability of being sampled.
3. All probe–object intersections must occur at random.

These prerequisites are met if one uses a disector probe to estimate the total object number.

D. C. Sterio and the Disector Principle

In 1984 D. C. Sterio published a paper in the *Journal of Microscopy* entitled, "The unbiased estimation of number and sizes of arbitrary particles using the disector." This paper resolved the corpuscle problem by using a straightforward approach that provided a novel and theoretically unbiased method for counting the total number of objects from tissue sections. The author of the paper named the technique *the disector principle.* Rather than counting the total number of objects directly, the disector principle provides a theoretically unbiased estimator of the expected numerical density (N_V) in a defined reference space. As discussed later, to obtain a theoretically unbiased estimate of total number, N, the estimate of N_V from the disector principle must be scaled to the reference space.

The disector principle overcomes the corpuscle problem by counting number of cells with a 3-D probe, rather than by counting the number of profiles with a 2-D sampling plane. Thus, rather than counting profiles *on* tissue sections, the disector counts objects *in* tissue sections. A further procedural step involves an unbiased counting frame introduced by Gundersen in 1977, which avoids another source of bias (e.g., the edge effect) that is problematic for all studies of object number using a geometric probe. The optical disector provided the approaches necessary to estimate N_V on thin optical sections and to scale the local estimates of N_V to the complete reference space. In 1987, the critical issue of "how many animals, how many samples?" was addressed by Gundersen and Jensen, who showed that if sample estimates are repeated at systematic-random locations throughout a reference space, the variation around the mean estimate diminishes rapidly. When this sampling is repeated in a small number of individuals, the sample estimate rapidly converges on the expected value for the parameter for the population.

The Disector Probe

The disector is a "virtual" 3-D counting probe, in the sense that it does not occupy matter. The area of the disector frame [a(frame)], reported in square micrometers after adjustment for magnification, defines the x and y dimensions of the probe. The height, h, of the disector, the z-axis distance in which objects are counted, is also reported in micrometers after adjustment for magnification. The product of [a(frame)] and (h) constitutes the volume of the 3-D disector probe, in volumetric units (cubic micrometers). In practice, the observer scans within this 3-D disector probe using one or more sets of two adjacent planes (disector pairs) and counts the number of objects within each disector pair. Because the total volume of tissue scanned in this manner is known, the disector method counts the true number of objects in a known volume of tissue, N_V.

Estimating the true number distribution requires that an equal opportunity exist for counting every object in a well-defined tissue of interest. A corollary to this prerequisite is that no opportunity should exist for counting objects more than once. The corpuscle problem described by Wicksell (see Chapter 5) showed that routine histological tissue sections do not meet these criteria; instead, certain objects possess unequal sampling probabilities because of morphological characteristics (larger size, unusual shape) and orientations (longer axis perpendicular to the direction of sampling). Disector counting overcomes these limitations by counting all objects *for the first time only once,* regardless of any morphology and orientation in the tissue. "For the first time only once" means that each object within the 3-D disector probe is counted only at the initial intersection between the scanning plane and the object; at subsequent intersections, the object is not recounted. For this reason larger and smaller objects, objects with more or less complex shapes, and objects with their axes in any orientation have the same probability of being counted, without further qualification.

The disector method described by Sterio uses an unbiased frame and counting rules to avoid edge effects. This term refers to the potential for overcounting along the edges of the 3-D disector probe. The consistent designation of half of the edges of the 3-D disector probe as exclusion surfaces and the other half as inclusion surfaces avoids edge effects. For example, the upper and right lines on the disector frame may be inclusion lines, while the lower and left lines may be exclusion lines; the top plane of the 3-D probe may be designated an inclusion plane, and the bottom plane an exclusion plane (Figure 6-1). A final feature of the 3-D disector probe is that the exclusion lines extend to infinity above and below each disector. As discussed later, these extensions ensure that irregularly shaped objects have the same probability of being counted as regularly shaped objects. Figure 6-2 shows cells (arrows) within the counting frame and not touching the exclusion lines.

In practice, counting rules drive counting decisions. Objects that touch either the inclusion planes or lines or fall within the disector frame are counted (included); objects touching the exclusion planes or lines or that fall outside the disector frame are not counted (excluded). Theoretically unbiased counting rules prevent the introduction of bias from edge effects and thus ensure that all objects have the same probability of being counted, regardless of their size, shape, and orientation.

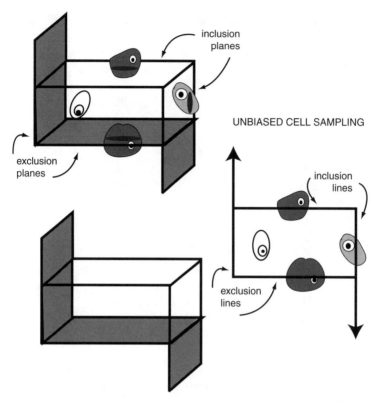

Figure 6-1. Unbiased disector counting frames showing inclusion and exclusion lines (2-D) and inclusion and exclusion planes (3-D)

The theoretically unbiased counting frame contains a modification that avoids a special type of edge effect, as shown in Figure 6-3. The biased frame is a simple square in 2-D and a simple cube in 3-D; the theoretically unbiased frame adds extensions from the upper left and lower right corners. Application of the unbiased counting rules to the irregular object profile shown in Figure 6-3 illustrates this simple yet effective modification. For the biased frame, the profile of the irregular object of interest hits the inclusion lines in two frames (5 and 8), and therefore has a higher probability of being counted than a profile with a less complex border. As shown in the unbiased frame, extending the exclusion lines beyond the upper left and lower right edges of the frame eliminates this unequal sampling probability; the profile hits the inclusion line in frame 8 and the exclusion line in frame 5. With the unbiased frame, all profiles regardless of their shape are counted by a single frame; thus, all profiles have the same probability of being counted.

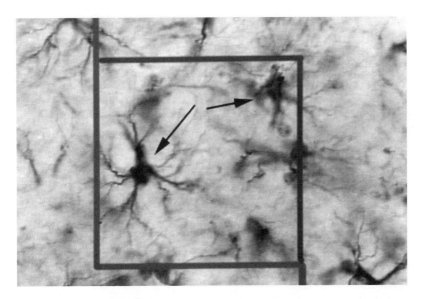

Figure 6-2. Unbiased counting rules are required to avoid bias from oversampling objects at the edges of geometric probes. Objects hitting inclusion lines or falling inside the frame are counted (arrows).

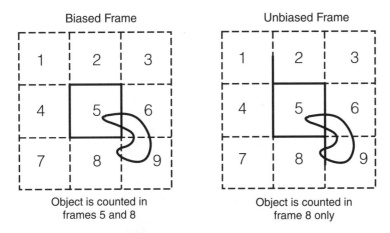

Figure 6-3. Biased versus unbiased counting frames

Guard Volume

A guard volume is a tissue space above and below the 3-D disector probe in which no counting is done. A guard volume avoids an edge effect that arises from the passage of the knife blade through tissue during preparation of sections. As the leading edge of the knife pushes through relatively soft tissue, objects in the path of the knife can be cut, torn from the tissue, or pushed out of the way either above or below the sectioning plane. This passage of the knife blade artificially changes the number of objects appearing at the cut surfaces of the tissue. Instead of counting cells at the surface of a section where the artifacts occur, the observer uses the guard volume to avoid these artifacts created at upper and lower surfaces of the section. The size of the guard volume varies slightly with changes in the section thickness. An optimal guard volume is one-fifth to one-fourth the expected diameter of the objects being counted, for example, 4–5 μm above and below the disector for a population of cells averaging 20 μm in diameter.

In Figure 6-4, a schematic of a tissue section viewed from the side shows the guard volume as a hatched area; the tops of cells are not counted in the guard volume. Cells are counted when the topmost point comes into focus while one is optically scanning through the tissue with thin focal planes.

Counting Cells

Bringing these concepts together enables one to make theoretically unbiased estimates of the total number of cells in a known volume of reference space. Beginning with a systematic-random set of sections through a reference volume of tissue containing the objects of interest, the observer outlines the reference space and selects a random x-y location to begin counting. If the objects contain unique subunits that are unambiguous and recognizable

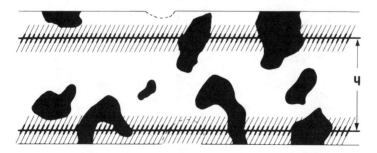

Figure 6-4. Schematic view of an optical disector frame. Hatched lines indicate guard volume to avoid artifacts at section surface, h, height of disector. (From Gundersen, 1986.)

(e.g., nucleus, nucleolus), these subunits may be used as the counting item for the object. A count of 1 (Q^-) is registered each time a new counting item appears in the disector volume.[2] The sum of new counting items is indicated by ΣQ^- in stereological equations. The theoretically unbiased counting rules provide the criteria for resolving decisions about which object to count without overcounting objects at the edges. Not all object profiles hit by the counting frame are counted; rather, the counting rules ensure that all objects regardless of their size, shape, and orientation have an equal probability of being counted.

In the typical efficient design, this counting process is repeated by moving the unbiased counting frame to between 100 and 200 systematic-random locations on 8 to 10 systematic-random sections through the entire reference space. The sum of the objects counted [ΣQ^-] divided by the total volume of the disector probes ($\Sigma\ Vol_{samp}$) provides a sample estimate of the total number of objects counted in a known volume of the reference space, also known as the numerical density, N_V:

$$N_V = \Sigma Q^- / \Sigma\ Vol_{samp}$$

where:

$$\Sigma\ Vol_{samp} = n \times Vol_{dis} = n \times [a(\text{frame}) \times h]$$

and

N_V	= estimate of total number per unit of volume (μm^{-3})
ΣQ^-	= sum of objects counted
$\Sigma\ Vol_{samp}$	= total volume of disector probes (μm^3)
n	= total number of disectors sampled
$a(\text{frame})$	= area of the disector frame (μm^2)
h	= height of the disector probe (μm)
Vol_{dis}	= volume of a single disector (μm^3) = $[a(\text{frame})h]$

Practical Applications of the Disector Principle

There are three primary applications of the disector principle: the physical disector, the optical disector, and the optical fractionator. The original formulation of the disector method was designed for scanning tissue using a pair of thin sections, a process that has come to be known as the physical disector. The approach was termed *physical* because the disector counting is done on actual (physical) sections. Today the physical disector is limited to count-

2. Q is from the German term *Querschnitt* meaning cross-section; the minus sign indicates that a cross-sectional profile disappears between the two sections of a disector pair.

ing objects on photographic images of sections too thin for optical scanning, for example, electron micrographs. The optical disector, first proposed by Gundersen in 1986, is an extremely efficient modification of the disector principle for counting objects on relatively thick sections. Finally, the optical fractionator is a combination of the optical disector and the fractionator sampling scheme, first proposed by West et al. in 1991. A brief description of each of these applications of the disector principle follows.

The Physical Disector

The physical disector is the preferred choice for estimating N_V on thin sections. Examples include estimation of N_V on pairs of adjacent electron photomicrographs, in a stack of confocal images, or for light microscope studies on adjacent thin sections. The distance separating the sections defines h, the height of the disector. Objects to be counted in the tissue are separated by two thin adjacent sections (a disector pair). One section of the pair is the "reference" section and the other is the "lookup" section (Figure 6-5). The critical factor when selecting the disector height is that the distance should be less than the smallest diameter of any object in the population. This minimal value for the disector height eliminates the possibility that an object

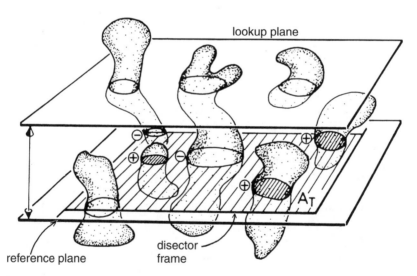

Figure 6-5. In 3-D a disector consists of a lookup plane (upper) and reference plane (lower) separated by a known distance, h, the disector height. Objects hitting the area of the counting frame on the reference plane are counted (+), if and only if the object is not present on the lookup frame and does not hit the exclusion lines. A_T, area of disector frame. (From Cruz-Orive and Weibel, 1990.)

could fall between the sections of the disector pair, thus appearing on neither the reference section nor the lookup section.

Using systematic-random sampling, the first field of view is located for the reference section. The same location is found on the lookup section, using structures in the tissue (e.g., blood vessels, tissue edges) for orientation. For use with the disector principle, a grid containing counting frames is placed at random over a tissue section containing the reference space of biological interest. Using the unbiased counting frame and counting rules to avoid edge effects, objects in a known volume are counted. Counts (Q^-) are registered when an item to be counted is present in the frame on the reference section, but is absent on the lookup section. This fits the rule mentioned earlier for counting each object in a known volume *for the first time only once*. No counts are registered if the same object appears on both sections of the disector pair. This procedure is repeated at 100 to 200 x-y locations for the entire reference space. The same approach is used for estimates of object number using disector pairs from electron micrographs or confocal images; there are two adjacent focal planes, one reference plane, and one lookup plane.

When the number of disector pairs is limited (e.g., as in electron microscopy studies), increasing ΣQ^- in a defined volume of reference space by increasing the size of the disector frame will increase the efficiency of the counting process. Another efficient approach is to reverse the roles of the reference and lookup sections in the same disector pair. Because the scanning plane is moving in a different direction, different objects at the top and bottom of the disector will be included in the count, depending on which plane is the reference section.

The Optical Disector: Counting Objects in Thick Sections

In some sense the corpuscle problem evolved in parallel with the development of the microscope, beginning with its invention in the late sixteenth century by the father-and-son team of Jans and Zacharias Janssen in Holland. In the past 400 years, the need for greater resolution has been answered by increasingly powerful optics, and tissue sections have become increasingly thinner to allow highly condensed light energy to pass through the tissue. During this process the light beam carries morphological information to objective lenses where it is magnified and focused onto cells in the human retina. Magnification with light microscopy ranges from several times (2 to 5×) to over a thousand times, with a maximum resolution limited only by the wavelength of photons. In the case of electron microscopy, magnetic condensers focus a beam of electrons onto ultrathin tissue sections. Electrons reflected off objects in the tissue carry ultrastructural information to a collector and eventually to our eyes. Therefore, thinner sections permit more information to reflect off objects in the tissue and eventually into the eyes of the observer.

Although thin sections increase the clarity and resolution of biological structures in tissue sections, they contribute heavily to the corpuscle problem. Soon after Sterio introduced the disector principle as a solution to the corpuscle problem, new approaches were proposed for using it to count cells in biological tissue. Originally the disector principle was applied as the physical disector to disector pairs of photomicrographs; later, specialized side-by-side microscopes were developed that projected images of the reference and lookup sections onto a table. Although it was less expensive than counting from disector pairs of photomicrographs, using the side-by-side projection microscopes required tedious searching within adjacent sections for tissue landmarks before counting could be done. Eventually the now obvious (in retrospect) solution became apparent: Combine the two microscopes into one, and use thin focal plane scanning to count cells *only the first time once* as they came into focus within a virtual stack of optical planes within the thickness of the section. With the addition of a microcator, a device for accurate measurement of stage movement in the z-axis, the technology was complete. The user could optically section through a known distance in the z-axis and count the number of objects that appeared for the first time. Thus the optical disector, a remarkably efficient and straightforward application of the disector principle for light microscopy, was created.

Figure 6-6 shows a series of optical sections for counting cells in a known volume of tissue. No cells are counted in the top section through the guard volume. As the focal plane is moved through the z-axis of the section, cells are counted when they first appear within the disector frame or hit the inclusion lines.

The optical disector uses thin focal plane optical scanning to count objects as they come into focus within the known volume of the disector. The section must be thick enough to permit inclusion of a disector of sufficient height to count the objects of interest. The actual height of the disector is determined by the size of the objects to be counted, in relation to the section thickness. Thus, objects approximately 10 μm in rough diameter and with a guard volume of 5 μm above and 5 μm below the disector would be contained within a thickness ranging from 20 to 25 μm. Of course this thickness refers to the section after final histological processing is complete. Tissue sections cut at an instrument setting of 50 μm routinely show a final postprocessing thickness of 20 μm or less. As a practical measure, the guard volume can be decreased as necessary to accommodate variation in section thickness. The user must remain aware, however, of the possible bias introduced by artifacts

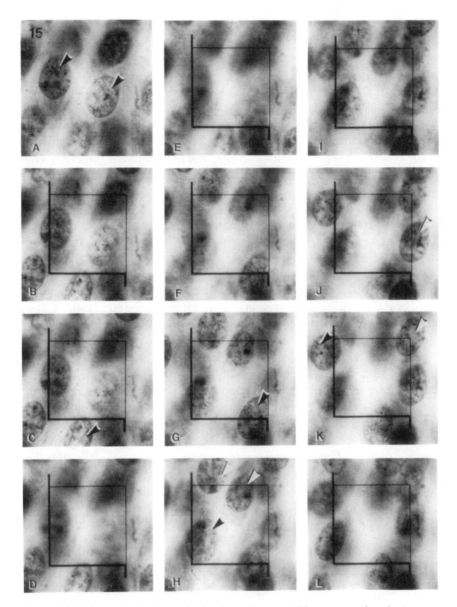

Figure 6-6. Cell counting in the z-axis showing cells counted (open arrows) and not counted (filled arrows). (From West and Gundersen, 1990.)

created at the section surfaces (e.g., lost caps) and not abandon the guard volume altogether.

Magnification

Although low-power objectives (5 to 40×) permit rapid visualization of wide regions for qualitative studies, these objectives are associated with relatively thick focal planes (depth of field). Counting objects by using the optical disector requires a thin focal plane for scanning the height of the disector in the z-axis. The depth of field of an objective lens limits the ability of an observer to identify structures at different vertical (z-axis) locations.

$$\text{depth of field} = [L(i_R^2 - NA^2)^{1/2}]/\ NA^2$$

where L is the wavelength of light (0.55 μm), i_R is the refractive index (about 1.52 for immersion oil, tissue, slide, cover glass, etc.), and NA is the numerical aperture. As shown in Table 6-1, as the numerical aperture of an objective lens increases, the depth of field decreases.

The optimal resolution for counting microscopic objects with the optical disector is the lowest oil-immersion objective in the range 60 to 100× that allows objects to be unambiguously identified. For counting exceedingly small objects (e.g., synapses and mitochondria) near the resolution limit of light microscopy, the thin focal planes afforded by a 100× oil immersion lens (NA 1.4) provide optimal resolving power and low depth of field. Naturally, for thin focal plane scanning of this type, the best optical settings (e.g., Köhler illumination) are required to capture all light refracted from the image on the stage.

The extraordinary increase in efficiency afforded by thin focal plane scanning gives the optical disector a great advantage over counting with the physical disector. The optical disector technique eliminates to a great extent the difficult, labor-intensive, and tedious effort required to precisely register adjacent sections of a disector pair using the physical disector. However, physical disector counting is the only option when optical sectioning is not

Table 6–1 Relation of Numerical Aperture to Depth of Field

Objective (x)	Numerical aperture	Depth of field (μm)
40	0.75	1.3
63	1.25	0.30
100	1.3	0.26
100	1.4	0.17

possible, for example, in electron microscopy, or if sections are cut too thin for optical plane scanning.

The Optical Fractionator

Of all the applications of the disector principle, the optical fractionator is the simplest to understand and easiest to use, provided the reference space can be exhaustively sectioned. In order to fully appreciate the advantages of this method, it is necessary to look at how estimates of local samples are scaled to obtain the total number of objects.

The disector method is theoretically unbiased for total N_V. However, because of the reference trap for density estimators, total N_V is biased for total object number, N. To avoid this potential bias, local density estimates of total N_V must be scaled to estimates of total number at the level of the reference space. The following sections discuss two methods for scaling local parameter estimates to the reference space: the two-stage approach and the fractionator method.

The Two-Stage Approach

The two-stage or $N_V \times V_{ref}$ approach is one method for scaling estimates of N_V to total object number. In this formulation, N is calculated as the product of N_V and total volume for the reference space, V_{ref}, which may be estimated by the Cavalieri method:

$$E(N) = N_V \times V_{ref} = [\Sigma Q^- \div \Sigma Vol_{dis}] \times V_{ref}$$

where:

$E(N)$ = estimate of total number of objects
N_V = numerical density (number per unit of volume)
ΣQ^- = sum of objects counted
V_{ref} = Cavalieri volume of reference space (volume units)
N_V = $\Sigma Q^- \div \Sigma Vol_{dis}$
ΣVol_{dis} = total disector volume = number of disectors $\times$ volume of one disector

Because V_{ref} and the denominator in N_V refer to a common volume, these units cancel in the product ($N_V \times V_{ref}$), leaving an estimate of the total number of objects, N. This cancellation eliminates changes in the reference space (shrinkage or expansion), resulting in a theoretically unbiased estimate of total N. The only caveat is that both N_V and V_{ref} must be measured on tissues that have had equivalent processing. When N_V and V_{ref} are measured on tissues that have received different processing, the result is a particular version of the reference trap. For instance, if V_{ref} is estimated on fresh tissue that is then fully processed (fixed, embedded or frozen, stained) before total N_V is estimated, changes in the reference space will occur between the estimation of N_V and V_{ref}. This differential shrinkage of the reference space will introduce bias into the product of $N_V \times V_{ref}$. This bias is easily avoided by estimating N_V and V_{ref} after final histological processing. In the next section we introduce a second method for estimating total number, the fractionator scheme. Among the primary advantages of the fractionator is that changes in the reference space during tissue processing do not affect estimates of total number, and thus by definition the method avoids the reference trap.

The Fractionator Scheme

According to Gundersen's seminal work in 1986, the fractionator scheme was originally developed to test the accuracy of the two-stage method. Gundersen astutely recognized that this calibration method provided a theoretically unbiased method for estimating the total number of objects in a defined reference space, and that this method was in some cases superior to the two-stage ($N_V \times V_{ref}$) method. Both methods scale local estimates of object numbers to the total reference space and thus avoid the bias associated with density estimates (the reference trap).

The fractionator scheme works as follows. The first step is to exhaustively section the reference space containing the objects of interest. From the total number of tissue sections containing any part of the reference space, a sample of sections is selected in a systematic-uniform-random manner. For instance, if the reference space produces 500 serial sections, to obtain 10 sections, a section is taken every 50 sections, with a random start between sections 1 and 50. For example, if the first section is number 31, then the sampled sections in the series will be numbers 31, 31 + 50 = 81, 81 + 50 = 131 . . . through the entire set of 500 sections. This fraction of the total number of sections containing the reference space (e.g., one tenth) is the section sampling fraction (*ssf*):

section sampling fraction (*ssf*) = number sections sampled ÷ total sections

The second step is to make a random sample of the total reference space on each section sampled in the first step. This step uses a theoretically unbiased counting frame as an areal sampling frame. The area of the counting frame [a(frame)] is spaced uniformly at x-y locations over and beyond the borders of the reference space on the surface of each section. The area between each x-y location defines the total area on each section (area x-y step). Thus, the area sampled on each section is the ratio of the frame area to the total area of the reference space on the section, that is, the area sampling fraction (*asf*):

area sampling fraction (*asf*) = a(frame) ÷ area x-y step

The third step is to take a random sample of the thickness of the tissue under each of the areas sampled in the second step. This step uses the disector to randomly sample tissue in the z-axis of the section. The thickness sampled in the section is a ratio of the frame height (h) divided by the section thickness (t), the thickness sampling fraction (*tsf*):

thickness sampling fraction (*tsf*) = $h ÷ t$

From this point on, the procedure is as described above for counting objects using the disector principle. The same counting rules apply: The number of objects is counted for the first time once in a known fraction of the total thickness, under a known fraction of the total area, on a known fraction of sections through the total reference volume.

$$E(N) = \Sigma Q^- \times F_1 \times F_2 \times F_3$$

where:

$E(N)$ = estimate of total number of objects
ΣQ^- = sum of objects counted using disector counting rules
F_1 = 1/section sampling fraction = 1/(number of sections sampled/total sections)
F_2 = 1/area sampling fraction = 1/(area of section sampled/total area)
F_3 = 1/thickness sampling fraction = 1/(disector height/section thickness)

Comparison of Two-Stage and Fractionator Methods for Total N

The fractionator approach is similar in two ways to the two-stage method: Both scale local estimates to the reference space and both generate local estimates using the disector principle. With regard to the first similarity, there are important differences in how local estimates are scaled to the reference space. First, because fractionator estimates of total number are not based on the reference volume (V_{ref}), the estimates of number are unaffected by artifactual changes in the reference space during tissue processing.

Second, because the fractionator method does not require an estimate of total volume of the reference space, there is no need for decisions about the borders of an ill-defined reference space, as required for the two-stage method. In practice, the reference space and surrounding tissue are covered during the process of x-y stepping over each section. At each x-y location, when an object falls within the disector, the observer simply asks "Is this the object I am counting?" If the answer is yes, the object is included in the count; if the answer is no, it is omitted. Notice that the decision process does not require defining the borders of a reference space per se, only that the objects of interest can be unambiguously recognized. The single potential disadvantage of the fractionator scheme is that the total reference volume must be exhaustively sectioned to obtain the denominator of the section sampling fraction (the total sections through the reference space). In most experimental studies, the reference space is easily reduced to serial sections. In these cases the effort to exhaustively section the reference space is inconsequential.

Counting in a Known Fraction

As can be seen from the preceding discussion, the final word on estimating total object number belongs to the optical fractionator. It was first introduced to count neurons in a defined brain region and enjoys the advantages of the optical disector and the fractionator sampling scheme. The reference space is exhaustively sectioned and then sampled according to the fractionator scheme. These sections are then used to determine the section sampling fraction, area sampling fraction, and thickness sampling fraction. At each x-y location in the sampling scheme, the 3-D optical disector and *first time only once* rule are used to obtain a theoretically unbiased estimate of objects in a known fraction of the reference space. Using the reciprocal of the sampling fractions, the number of objects is multiplied by the product of these reciprocal fractions to estimate the total number of objects for the total reference space.

Summary

Since the invention of the microscope, biologists have struggled with the problem of counting the total number of objects in a defined region of tissue, based on the number of object profiles on tissue sections. From the work of Delesse we know that the number of profiles per unit of reference area is not equal to the number of objects per unit of reference volume. In the 1920s Wicksell articulated the corpuscle problem as sampling bias resulting from different probabilities of sampling biological objects based on their size, shape, and orientation relative to the direction of sectioning.

Finally in 1984, Sterio proposed the disector principle to solve the corpuscle problem, and in doing so provided the first theoretically unbiased estimator of total number of objects for a defined reference space of biological tissue. Disector counting uses counting rules and other techniques to avoid the introduction of bias from edge effects.

Later, Gundersen showed that local estimates of total N_V could be scaled to the total reference space using the two-stage ($N_V \times V_{ref}$) and fractionator methods, and thus avoid the reference trap associated with density-based estimates of object number. Gundersen and his colleagues in Denmark then adapted these methods to the use of thin focal plane scanning of thick sections with the optical disector and optical fractionator methods (see Gundersen, 1986 and West et al., 1991).

7

Volume

This chapter discusses theoretically unbiased estimators and geometric probes for estimating the volume of nonclassically (arbitrarily) shaped biological structures. Volume is a parameter of significant biological importance. In anatomical regions and populations of cells there is variation in volume (i.e., biological variation); differences in volume occur normally over time (e.g., normal aging) and differences also result from pathological, immunological, and toxicological conditions (atrophy, hypertrophy). During normal cell division, the volume of cell nuclei increases, and in some disease conditions, such as certain cancers, the mean nuclear volume of cells in the affected tissue also increases; in other cancers, mean nuclear volume is reduced. An understanding of the significance of differences in volume can provide insight into the underlying cellular processes. Methods for quantifying the volume of nonclassically shaped objects date back to third-century Greece and Archimedes' method of determining volume by water displacement. Since then a number of sterological methods have been developed, or in some cases adapted from other disciplines, for extracting accurate volume information for arbitrarily shaped 3-D objects based on the area of their arbitrarily shaped profiles on 2-D sections.

Volume is a major stereological parameter of interest in a wide variety of biological studies. Volumetric differences are used in research on development and aging, pathological conditions, and experimental manipulations. They are also important in comparative studies of vertebrate and invertebrate anatomy and the relation between structure and function. For example, the cerebral cortex of the primate brain has undergone a rapid and relatively large expansion in volume compared with other brain regions. This remarkable expansion occurred in concert with an increase in function—a rise in the cognitive ability of primates.

On the other hand, studies of Alzheimer's disease have shown that the severity of dementia is strongly correlated with a reduction in the total volume of the cerebral cortex. In addition to macroscopic changes in brain anatomy, neuropathology involves volumetric changes at the microscopic level. In a number of neurodegenerative diseases associated with loss of voluntary control of the motor system, including amyotrophic lateral sclerosis (Lou Gehrig's disease), Parkinson's disease, and Huntington's chorea, there is loss of volume and cell death in regions of the nervous system that underlie motor control.

A further example of structure–function correlates involving volume is neointimal thickening, the pathology in the lumen of coronary arteries that supply blood to the heart. Following certain procedures, for example, heart transplantation, increases in the volume of the neointima reduce arterial blood flow, eventually leading to rejection of transplanted hearts. Finally, there are several cancers, including malignant melanoma and bladder cancer, in which increases in the nuclear volume of cells are a reliable predictor of metastasis, that is, spread to new areas of the body. To correctly correlate these and other morphological changes with the underlying functional end points, accurate methods must be used to quantify the volume of defined regions of biological tissue and defined cell populations at the macroscopic and microscopic levels.

In this chapter we discuss the estimators and the probe that generate theoretically unbiased estimates of volume (size). Because volume is a 3-D parameter, estimates of volume using geometric probes can vary from zero to three dimensions, allowing greater flexibility in the geometric probes used to quantify volume (point, lines, planes) than those used with number, a zero-dimensional parameter. Stereological estimators of volume can be divided into methods to estimate an entire reference volume, known as regional volume estimators, and those that estimate the mean volume of an object, also known as local size estimators.

Reference Spaces: Volumes in 3-D, Areas in 2-D

Before using stereology to estimate a parameter, we must first identify the structure of interest. As indicated earlier, the bounded biological space of interest is the reference space (reference volume). As discussed in Chapter 2, a reference space refers to an anatomically and functionally significant volume of tissue in 3-D. In biological applications we are frequently faced with reference volumes that appear as reference areas on 2-D tissue sections. As we have seen, classical geometry provides numerous estimators for volume, but these models are generally not appropriate for biological objects. For example, the model-based estimator for the volume of a sphere with radius, r, $V_{sphere} = 4\pi r^3/3$, is accurate only for spheres. Thus the challenge is to estimate the 3-D volume of biological objects from the reference areas on 2-D sections without relying on inappropriate formulas and correction factors based on nonverifiable models and assumptions.

Archimedes and the Crown of King Hiero

The well-known account of Archimedes' discovery of the principle of volume displacement illustrates an early application of good stereological procedures to a nonclassically shaped object. Archimedes (287–212 B.C.) was well versed in the theoretical and practical applications of Euclid's classical geometry. Around the middle of the second century B.C., King Hiero of Syracusa requested Archimedes' help in resolving a problem: How could one determine the volume of an arbitrarily shaped crown? The king had promised Syracusa that a gold crown would be offered to the gods and gave a measured weight of gold to a royal goldsmith to complete the project. The goldsmith produced a crown that was a curving, delicate composition of grapes, intertwined vines, and fine branches of variable size, shape, and orientation. However, a rumor began circulating that the goldsmith had mixed a quantity of silver in the crown. Archimedes knew that the only way to confirm that the crown was pure gold was to determine its density (weight per unit of volume), and compare this with that of pure gold. However, the approaches available to Archimedes, namely, Euclid's model- and assumption-based geometry, offered no methods for measuring the volume of a nonclassically shaped object. From Vitruvius' account: "While Archimedes was considering, he went down into the bathing pool. There he observed that the amount of water which flowed outside the pool was equal to the amount of his body that was immersed. Since this fact indicated the method of explaining the case, he did not linger, but moving with delight he leapt out of the pool, and going home naked, cried aloud that had found exactly what he was seeking. For as he shouted in Greek: Heureka! heureka!"[3]

Archimedes discovered a way to determine volume by the displacement of water, which he subsequently used to find the volumes of the crown and an equal weight of pure gold, and compare their densities. From the viewpoint of good stereology, Archimedes' approach succeeded in large part because he avoided assumptions and models related to the shape of the crown.

---

3. Vitruvius Pollio, *The Ten Books on Architecture.* English translation by Morris Hicky Morgan, Oxford University Press, London, 1914, page 3. Citation provided from the Perseus database at Tufts University (*www.perseus.tufts.edu*).

About 2000 years later, the Italian mathematician Buonaventura Cavalieri, a student of Galileo Galilei, developed a more practical, though equally assumption- and model-free method for finding the total volume of nonclassically shaped objects.

The Cavalieri Method for Estimating Volume

Bonaventura Cavalieri (1598–1647) made his contribution to volume estimation in a work called *Geometria Indivisibilibus,* which is considered an important prelude to integral calculus. In an early demonstration of his principle, Cavalieri summarized the theory of indivisibles as follows: A line is made up of points as is a string of beads; a plane is made up of lines as is a cloth of threads; a solid is made up of planes as is a book of pages.

As a method to quantify the volume of an arbitrarily shaped solid, the Cavalieri principle falls historically between the methods of exhaustion used by the ancient Greeks and the integral calculus developed at the beginning of the eighteenth century.[4] From the viewpoint of applied stereology, Cavalieri showed that the volume of an arbitrarily shaped object can be estimated in an unbiased manner from the product of the distance between planes (T) and the sum of areas on systematic-random sections through the object (ΣA).

Cavalieri published his work in 1635, which today is known simply as the Cavalieri principle. It provided a practical alternative to Archimedes' method of estimating volume by displacement. Archimedes' method is not effective for estimating the volume of objects that absorb water rather than displace it, and it cannot easily be used to estimate the volume of different reference spaces within an object. Cavalieri's method showed that the volume of a population of objects could be estimated from the profile areas of the objects on cut sections. In practice, the Cavalieri approach requires an initial random cut through the reference space of interest, with subsequent cuts at consistent intervals, that is, systematic-uniform-random sampling (Figure 7-1).

This approach allows volume to be estimated by integrating the areas of a defined reference space. Provided the sections through the reference space are systematic-random, that is, all sections through the reference space have an equal probability of being sampled, the Cavalieri method gives an unbiased estimate of total volume. By repeating the estimate on several individuals from the population of interest, one obtains an estimate of the mean total volume for the reference space of interest. The geometric probe in this case

---

4. W. W. Rouse Ball, *A Short Account of the History of Mathematics,* 4th ed., 1908, as transcribed by D. R. Wilkins, School of Mathematics, Trinity College, Dublin (www.maths .tcd.ie).

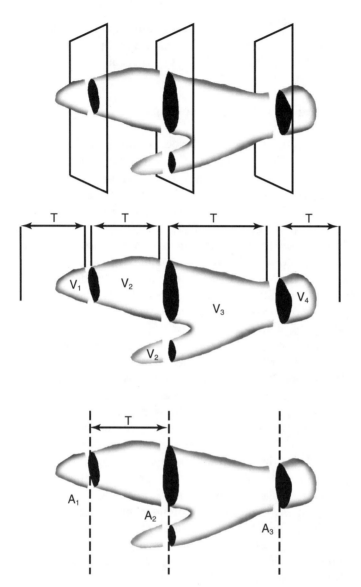

Figure 7-1. Schematic representation of sampling for estimating the total volume of an object using the Cavalieri principle

is a 2-D plane; the estimator is the Cavalieri principle. Several methods exist for estimating the areas on each of the cut surfaces of sections. One of the most efficient methods is point counting, which is described in greater detail later.

According to the Cavalieri estimator, the total volume of a reference space, V_{ref}, is equal to the sum of the reference areas on the cut surfaces of systematic-random sections through the reference space. The formula for estimating the Cavalieri volume is the product of the distance between the slices and the sum of the areas on the cut surfaces of each section:

$$V_{ref} = T \times (A_1 + A_2 + A_3 + A_4 + \ldots A_m)$$
$$= T \times (\Sigma A_{1-m})$$

where A is the area on the face of all sections 1 to m, ΣA_{1-m} is the sum of reference areas on the face of sections 1 to m, and T is the distance between each section.

Note that this theoretical unbiasedness of the V_{ref} estimate using the Cavalier estimator depends on the systematic-random sampling of the reference space. The remarkable efficiency of this approach in modern stereology stems from point counting, which developed from the work of several generations of geologists beginning more than two centuries after the Cavalieri principle was introduced.

Point Counting Methods

From Delesse to Point Counting

As mentioned in earlier chapters, Delesse showed that a particular phase on random sections cut in a rock is not related to the number of phase deposits in the rock ($N_A \neq N_V$; Chapter 5); however, the area of each phase on random surfaces cut through the rocks is proportional to the 3-D volume of each phase in the rock $A_{obj}/A_{ref} \approx V_{obj}/V_{ref}$. A half a century after Delesse's studies, another geologist named Rosival improved on Delesse's "cut and weigh" method for estimating volume fraction from area fraction. Rosival showed that the length of randomly placed line segments hitting profile areas is proportional to the object's volume in 3-D. Expressed as fractions of the total reference area, these equalities are $L_{obj}/L_{ref} \approx A_{obj}/A_{ref} \approx V_{obj}/V_{ref}$.

In the 1930s, two other geologists, Thomson in 1930 and then Glagolev in 1933, showed that Rosival's method was unnecessarily complicated. One need only place a grid of points over the cut surface; provided the point grid is placed at random with respect to the profiles, the number of points hitting the profile is proportional to the volume of the object in 3-D. Thus, the sum of the points hitting the object profiles (i.e., the point fraction, P_{obj}/P_{ref}) is proportional to the line fraction, the area fraction, and the volume fraction of the objects in 3-D: $P_{obj}/P_{ref} \approx L_{obj}/L_{ref} \approx A_{obj}/A_{ref} \approx V_{obj}/V_{ref}$ (see Figure 7-2).

Random grid placement ensures that only the parameter of interest in-

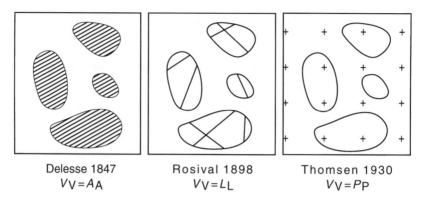

Delesse 1847	Rosival 1898	Thomsen 1930
$V_V = A_A$	$V_V = L_L$	$V_V = P_P$

Figure 7-2. The development of point counting to estimate profile area

fluences the probability of a probe–object intersection. In the next section we use probability to show that the total number of points hitting area profiles is proportional to the profile area, without further assumptions.

The Accuracy of Point Counting

A point grid is a geometric probe of known dimensions; the area per point, a(point) is the product of the distances between points in the x- and y-directions. On each section containing a reference area, the point grid is placed at random over the profiles of interest. The number of points (P) hitting the profiles is counted. According to Thomson and Glagolev, the total area of the profiles is the product of the sum of the points (ΣP) hitting the cut surface and a constant, the area per point $[a(p)]$ on the point grid. Thus, for point counting to provide an estimate of profile area, the sum of the points hitting profiles must be proportional to the area of the profiles, with no further qualification.

Because the point itself has a nonzero area, we must avoid introducing bias from points of different sizes. Thus, we select a dimensionless location associated with each point to act as the counting item. Any location on the point will work (e.g., where the upper and right arms of the point converge) (arrow in Figure 7-3), provided the same location is used consistently. When this "point within the point" hits the reference area, the point is counted; otherwise, it is not counted. The reference area, α, is indicated by the hatched region in Figure 7-3. A point grid of known area per point is placed at random over the reference space containing the profile of interest, A. The estimator (*est*) for the expected value of α, the area of A, is

$$est\ \alpha = \Sigma P \times a(p)$$

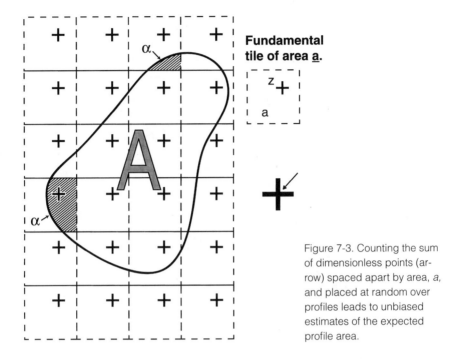

Figure 7-3. Counting the sum of dimensionless points (arrow) spaced apart by area, *a*, and placed at random over profiles leads to unbiased estimates of the expected profile area.

where $a(p)$ is the area per point = distance in the x-direction × distance in the y-direction and ΣP is the sum of points intersecting A at random.

What is the probability that the estimator, $\Sigma P \times a(\text{point})$, is a theoretically unbiased estimator of α?

Set $P = 1$ if z intersects A
Set $P = 0$ if otherwise

Possible outcomes:

P	Probability	$est\ \alpha = P \times a(p)$
1	α/a	a
0	$1 - \alpha/\alpha$	0

Adding the probabilities for the two possible outcomes,

$$est\ \alpha = [(\alpha\ /\ a) \times a] + [1-(\alpha\ /\ a) \times 0] = \alpha$$

The estimator $[a(p) \times \Sigma P]$ shows that the estimate of α equals the expected value for α, the expected area of A. Thus, the number of points on a point

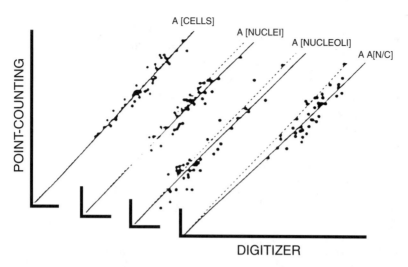

Figure 7-4. Comparison of point counting and pixel counting (digitizer) to assess area parameters for a defined population of cells. (From Gundersen et al., 1981.)

grid of known area per point is a theoretically unbiased probe for estimating the area of an arbitrarily shaped profile, with no further qualification.

In 1981 Gundersen et al. showed that point counting was an efficient alternative to estimating area by pixel counting. Their study quantified the profile areas of cells, nuclei, and nucleoli using a conventional pixel-counting method (digitizer) and point counting (Figure 7-4). The results showed that both methods could give similar results. However, the two approaches differed markedly in precision and efficiency, with point counting being superior to pixel counting for areal estimation. The reason for the difference was attributed to the time needed to achieve a precise estimate. The pixel-counting approach showed markedly lower levels of precision (higher variation) unless time and effort were taken to carefully outline the structures of interest; second, pixel counting required a time-consuming editing process to exclude the areas of extraneous structures from the estimate.

As shown in the next section, point counting provides a theoretically unbiased, practical, and efficient geometric probe (point grid) for estimating the total volume of an arbitrarily shaped reference space. Five decades after point counting, and over 350 years after the Cavalieri volume estimator was proposed, Gundersen and Jensen showed in 1987 that the combination of geometric probes and the Cavalieri estimator provides a theoretically unbiased and efficient method for estimating total volume from systematic-random sections through arbitrarily shaped objects.

The Cavalieri-Point Counting Method

The first step in the Cavalieri-point counting method is to make systematic-random sections through the reference space. A uniform array (grid) of points with a known area per point $[a(p)]$ is placed at random over the reference areas on the cut surfaces of the sections, with each section taken in serial order from the first to the last section containing the reference space. The observer counts the number of points hitting the reference area on each section. To quantify the volume of multiple structures in the tissue simultaneously, the number of points hitting each structure is tallied separately. Mathematically,

$$A_{obj} = \Sigma P_{obj} \times a(p)$$

where ΣP_{obj} is the sum of points (P) hitting the profile of interest and A_{obj} is the reference area.

For magnified images, the observed $[a(p)]$ is corrected by the areal magnification $[(mag)^2]$: corrected area = $a(point)/(mag)^2$. For a systematic-random set of sections through a reference space, random placement of the point grid followed by counting the sum of points hitting object profiles is repeated on each section, in serial order, through the entire reference space. Because the Cavalieri estimator is unbiased for volume and point counting is unbiased for area, the Cavalieri-point counting method provides a theoretically unbiased estimate of the expected volume (Figure 7-5). To test the accuracy of the Cavalieri-point counting method, we compared this method with the Archimedes principle to estimate volume for defined regions of biological tissue, as shown in Figure 7-6.

The intra-rater and inter-rater reliability of the Cavalieri-point counting method is shown in Figures 7-7 and 7-8. Figure 7-7 shows the results for two operators estimating volumes for the same main region (the cortex); Figure 7-8 shows the results for the same operator making four repetitions of volume estimation on the same brain.

For reference spaces of biological interest, the Cavalieri-point counting method provides the most efficient and easy-to-use approach for estimating 3-D reference volumes from tissue sections. It applies well to tissue sections and can be easily applied to multiple reference spaces within a larger reference space. It estimates an absolute parameter (volume), not a ratio (volume fraction), and thus requires no assumptions about changes in the reference space during tissue processing. Theoretically unbiased and efficient volume estimates can be obtained from no more than 8 to 10 systematic-random sections through a reference space. Across these sections, one should count 100 to 200 total points spaced in a systematic-random manner. This number of

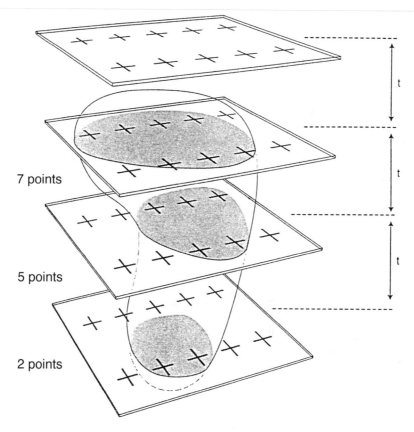

7 points

5 points

2 points

Figure 7-5. Combination of Cavalieri volume and point counting to estimate total volume of an arbitrarily shaped object

sections and points analyzed will capture most of the within-section and between-section variation in any reference space. When more than 8 to 10 sections are present, one can simply subsample this number from the total number of sections; when more than 200 points hit the reference space on all sections, the spacing between points on the point grid is simply increased. Because the greatest source of variation in essentially all stereological estimates is the variation between individuals, the optimal strategy in terms of efficiency is to sample lightly within individuals and to focus effort and resources on sampling enough individuals to make a stable estimate for the population.

Thus, 2300 years after Archimedes' first effort, the Cavalieri-point count-

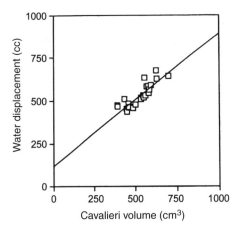

Line formula: y=0.774x +120.451 r=0.855

Figure 7-6. Strong, statistically significant correlation between total volume estimates for human brain cortex using Cavalieri-point counting and Archimedes' water displacement methods. (From Subbiah et al., 1996.)

ing method remains the method of choice for estimating volumes of biological objects cut into 2-D tissue sections. Chapter 11 provides a detailed example of the use of the Cavalieri-point counting method to estimate total volume for a defined biological reference space.

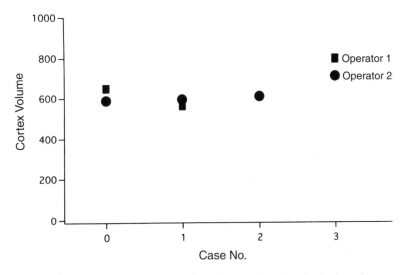

Figure 7-7. Strong inter-rater reliability for estimates of total cortical volume for four human brains

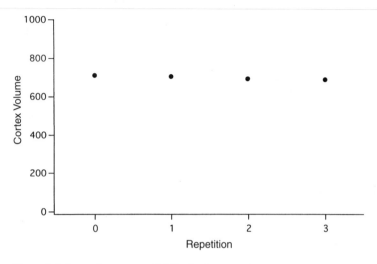

Figure 7-8. Strong intra-rater reliability for four repetitions to estimate total volume of human brain cortex

Comparison of Methods for Estimating Volume

The Cavalieri approach uses a 2-D plane as a geometric probe to sample and estimate the volume of any reference space, regardless of its size, shape, or orientation. Similarly, the Archimedes principle measures volume as a function of the volume of water displaced. Like Archimedes' method, Cavalieri's method is theoretically unbiased in the sense that the expected value of the sample estimate equals the expected value for the parameter. It is useful, however, to observe how sensitivity differs between Archimedes' principle and Cavalieri's principle.

The Archimedes principle allows one to determine volume using a calculation. For an extremely rough and quick estimate, one could use a graduated cylinder calibrated in milliliters; a more finely calibrated cylinder would give a measurement of greater sensitivity. Both methods are theoretically unbiased because both are based on a theoretically unbiased concept, the Archimedes principle. The sensitivity of an individual V_{ref} estimate made by the Archimedes method depends on how finely the device used to measure the volume of displaced water is calibrated.

Compare the measurement of volume using the Archimedes principle with the estimation of volume using the Cavalieri method. The latter approach involves a number of estimates based on stochastic geometry and probability theory. That is, provided a random first cut is made through each object in the analysis, the sum of the areas of the reference space on the cut surface of each section is proportional to the expected value. Repeated sampling of sections through the reference space will converge the sample estimate on the population parameter. Continuing to sample a greater number of sections initially increases the precision of the estimate, but a point of diminishing returns is rapidly reached. At this point, which is usually between 8 and 10 systematic-random sections, adding more systematic-random sections to the analysis leads to only minor increases in precision of the sample estimate. The reason is that most of the within- sample variation in total volume can be captured with 8 to 10 sections. As discussed in Chapter 10, the level of sampling within each individual should be based on the biological variation in the volume of the reference space in question.

Thus, the goal of the Cavalieri principle is not to determine V_{ref} for the reference space in a single individual, but rather, to estimate the mean V_{ref} for the population of individuals. As for all stereological estimates, theoretical unbiasedness refers to the mean sample estimate of V_{ref}, rather than the estimate of V_{ref} in any one individual. In this case, the accuracy of the estimate increases as the sample estimate of V_{ref} approaches the expected value, μ, for the population.

In addition to these theoretical differences, there are further practical advantages of the Cavalieri approach over the Archimedes principle. The Cavalieri method is easily applied to objects cut into slices, thus lending itself to the analysis of defined reference spaces on cut sections of biological tissue. Archimedes' principle, in contrast, is limited to nonporous objects that can be fully immersed in water. A second practical advantage of the Cavalieri method is that it can be used to estimate mean V_{ref} for separate reference spaces within a larger volume. The Archimedes method is difficult to apply to smaller reference spaces contained within a larger reference space.

Unlike the Archimedes and Cavalieri estimators, both of which estimate a first-order stereological parameter (total volume), the Delesse method estimates a ratio, volume fraction (V_{obj}/V_{ref}). As for all ratio estimators, there is a major assumption inherent in applying the Delesse principle to biological tissue: The relationship $A_{obj}/A_{ref} = V_{obj}/V_{ref}$ is only true under a critical assumption: that the area and volume of the object change in direct proportion to the area and volume of the reference space. If the denominators in the formula change independently of the numerators, then the relationship $A_{obj}/A_{ref} = V_{obj}/V_{ref}$ is no longer valid.

Local Size Estimators

We have shown that the unbiased Cavalieri estimator with points as the geometric probe provides perhaps the most practical and efficient method for es-

Corresponding Sphere of Equivalent Area
$4\pi r^3/3$

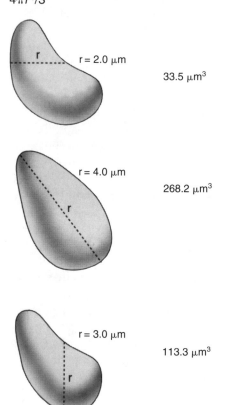

r = 2.0 μm

33.5 μm³

r = 4.0 μm

268.2 μm³

r = 3.0 μm

113.3 μm³

Figure 7-9. Potential bias associated with the assumption of sphericity for estimating total volume of anisotropic objects

timating the total volume of defined reference spaces on tissue sections. For volume estimates of objects too small to quantify using a plane through the object, a variety of unbiased size estimators have been developed in the past two decades, beginning with the nucleator principle. The importance of using theoretically unbiased methods is that wide variation exists when using assumption-based methods to estimate mean object volume, as shown in Figure 7-9.

The Nucleator

The nucleator principle is based on the idea that a mean line length obtained from randomly oriented lines across profiles of a population of objects is a theoretically unbiased estimator of the mean volume for the objects.

The nucleator estimator is used in conjunction with 1-D line probes. The theoretical basis for this estimator can be understood in terms of the geometry we learned in high school.

A radius, r, can be defined as a straight line from the center of a circle to the border. When one uses a radius as a geometric probe for area, the area of the circle can be calculated from the formula $A = \pi r^2$. When applied to 3-D, a radius can be used to calculate the volume of a sphere from the formula $V = 4\pi r^3/3$. In applying stereology to biological objects, we expand these formulas to general relationships that permit theoretically unbiased area and volume estimates for objects of all shapes and dimensions, not only those composed of circles and spheres. For straight lines oriented in a random direction on 2-D profiles through a population of 3-D objects, the mean line length across the profile is proportional to the mean volume of the objects. To convert from a proportionality to a parameter estimate, we find the mean value of the cube line lengths, mean l^3. Finally, we convert these area estimates to the mean volume of the objects according to the nucleator estimator: mean object volume = $(4\pi/3) \times$ mean l^3 (see Figure 7-10).

As shown in Figure 7-11, the nucleator applies to all profiles, regardless of shape. The fiduciary point chosen within the profile does not affect the accuracy of the estimator. The critical factor is that the direction from the point to the border must be random. For cells, the nucleus tends to be located toward the center of a profile, rather than eccentrically. Points consistently chosen near the nucleus will introduce less variation in the estimate of mean cell volume than points chosen within the profile at random; hence, the estimator is named the nucleator.

The nucleator method is unbiased for the expected mean volume of objects, provided the isotropy requirement is met: all possible integral directions for l have equal probability. To ensure isotropy, either the direction (angle) for l must be randomized on the surface of each profile or the orientation of the object must be randomized in 3-D prior to sectioning the cells within the tissue into profiles. Either approach ensures that all possible line lengths have equal sampling probability. Of course, if a population of cells were inherently isotropic, that is, if the cells were spheres, then theoretically unbiased estimates of mean object volume could be easily determined on sections cut at any convenient angle (e.g., coronal). Because populations of naturally occurring cells are not isotropic, estimating the mean volume of objects using this assumption introduces an unknown, unmeasurable, and unremovable quantity of bias into sample estimates. Instead, cells must be rendered isotropic using vertical-uniform-random (VUR) sections or isotropic-uniform-random (IUR) sections.

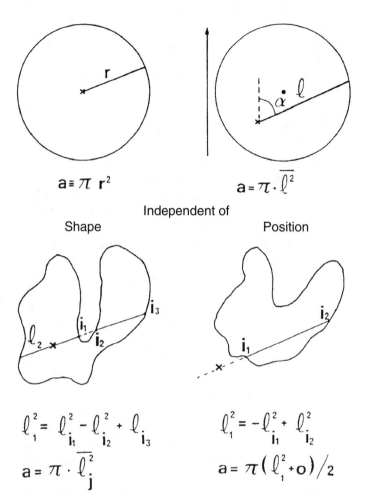

$$a \equiv \pi \, r^2$$

$$a = \pi \cdot \overline{\ell}^2$$

Independent of

Shape   **Position**

$$\ell_1^2 = \ell_{i_1}^2 - \ell_{i_2}^2 + \ell_{i_3}$$

$$a = \pi \cdot \overline{\ell}_j^2$$

$$\ell_1^2 = -\ell_{i_1}^2 + \ell_{i_2}^2$$

$$a = \pi \left(\ell_1^2 + o \right) \big/ 2$$

Figure 7-10. The nucleator principle. The formula for the area of a circle based on the radius ($A = \pi r^2$) is derived from the more general relationship for estimating volume from the mean line length l from any reference point within a profile to the border in a random direction; that is, $A = \pi \times$ mean l^2, if and only if the angle α is unbiased between 0 and 2π. In contrast to the specific formula for circle area ($A = \pi r^2$), the general relationship holds for all profile shapes and positions of the reference point relative to the profile. (From Gundersen et al., 1988.)

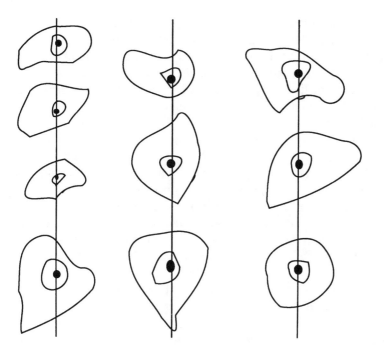

Figure 7-11. A vertical axis of rotation for arbitrarily shaped objects

The Nucleator and Isotropic-Uniform-Random Sections

Randomization of object orientation in 3-D can be guaranteed through the use of isotropic-uniform-random sections. In practice, for IUR sections, the entire reference space containing the objects of interest is dissected away from the surrounding tissue; the resulting tissue block is embedded in a random orientation prior to sectioning. After sections are stained to reveal the profiles of interest, a line grid is placed over the profile and the line length, l, is measured from the nucleus (or any point within the profile) to the border of the cell in a random direction. That is, a random angle is selected and a line is drawn from the nucleus to the cell border. To increase the efficiency of the method, after the first line length is measured, because the first angle is random, a second line length 180° from the first line can be measured. The mean of the two line lengths becomes the mean line length for a given profile.

The nucleator method randomizes the orientation of object profiles prior to measurement using randomly oriented line probes. However, we have not yet mentioned an important practical point: How does one select the profiles for measurement of mean line length? That is, on a given section there may be hundreds or thousands of object profiles. If we are interested in the mean volume for all cells in the population, then we will use a *number-weighted* approach (e.g., the disector principle) to sample and count cells before measuring their mean line length. That is, after a cell has been sampled and counted using the disector principle, a mean line length is determined. The procedure is repeated until we have estimated the mean line length for about 100 to 150 object profiles.[5] Using the disector principle to sample object profiles for measurement of mean line length generates an estimate of number-weighted mean object volume. That is, the estimate of mean object volume is related to the number distribution for the objects of interest.

Rather than a number-weighted sample of objects for measurement of mean line length, we may be interested in estimating the mean object volume for a particular subset of objects, for instance, the population of objects with relatively greater volume. In this case, we use a volume-weighted probe, a point grid, to preferentially sample the mean object volume for a population of objects with greater volume; we generate a *volume-weighted* estimate of mean object volume. In this case a point grid is laid at random over the cut surface of tissue containing the objects of interest. When a point hits an object's profile, that profile is sampled for inclusion in the estimate of mean object volume. This sampling approach is known as point-sampled intercepts, and will be revisited later in this chapter.

For both number-weighted and volume-weighted estimates of mean object volume, the same procedure is used for converting a set of mean line lengths into an estimate of mean object volume:

1. The line length, l, between the nucleus (or some other arbitrary point within the object's profile) and the border of the object's profile are measured in a random direction. If line lengths are measured in more than one random direction for the same profile, the separate line lengths are averaged into a mean line length for a given profile.
2. Each individual (or mean) line length for a given object profile is cubed, l^3.
3. The mean cubed line length, mean l^3, is calculated as the mean of the cubed line lengths for all profiles.

---

5. As detailed in Chapter 10, between 100 and 200 measurements within each reference space are sufficient to capture the biological variability for the majority of parameters of biological interest. Because the nucleator uses the nucleus to reduce the variability in estimates of mean line lengths and hence mean object volume, the optimal intensity of sampling turns out to be closer to 100 to 150 measurements per individual reference space.

4. The mean l^3 is converted to mean cell volume using the nucleator estimator:

mean cell volume = mean l^3 $(4\pi/3)$

where mean cell volume is the mean object volume, mean l^3 is the mean of the cubed line lengths, and $4\pi/3$ is a constant to convert area estimates into volume estimates.

As mentioned earlier, there are two options for ensuring isotropic intersections between the profile and the line probe. The first method uses IUR sections as described earlier. The second method, the rotator, uses vertical-uniform-random sections to estimate mean object volume based on the Pappus–Guldinius theorem (see the 1993 paper by Jensen Vedel and Gundersen).

The Rotator

In 1993 Jensen Vedel and Gundersen published an efficient modification of the nucleator method called the rotator. The rotator method is based on the Pappus–Guldinius theorem derived in part from work in the seventeenth century: "If a planar figure revolves about an external axis in its plane, the volume of the solid so generated is equal to the product of the area of the figure and the distance traveled by the center of gravity of the figure."[6] The rotator method can be used in conjunction with IUR sections, as described for the nucleator (i.e., the isotropic rotator). However, in its most efficient form, the rotator method is used in conjunction with vertical-uniform-random sections (the vertical rotator).

For VUR sections, tissue containing the objects of biological interest is blocked from surrounding tissue and rotated around an arbitrarily selected axis called the vertical axis. For maximum efficiency, the selection of the rotation axis should be the predominant long axis of cells in the tissue (Figure 7-11). After rotation around the vertical axis, the direction of sectioning is perpendicular to the vertical axis, that is, each section is parallel to the vertical axis.

After sections are stained to reveal the objects of interest, we again have the option to estimate the mean volume of the object from the number distribution (number-weighted mean object volume) or the volume distribution (volume-weighted mean object volume). In the former case, we sample objects for volume estimation using the disector principle; for the latter, we sample the object profiles using a point grid. Once an object profile is sampled, the vertical axis is identified and a line indicating the vertical axis of ro-

6. Quoted in E. B. Jensen Vedel and H. J. G. Gundersen, The rotator, *Journal of Microscopy*, 164: 21–27, 1993.

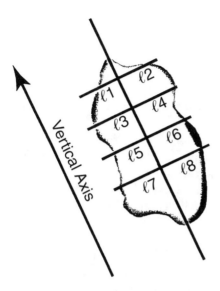

Figure 7-12. The rotator-vertical section method using four lines generates eight systematic-uniform-random line lengths (l_1 to l_8) on each object

tation is centered through the object at the nucleus. A grid of three or four lines is oriented perpendicular to the line indicating the vertical axis; the distance between the lines is systematic-random through the long axis of the object. That is, the first line is located at a random distance from the top of the object's profile, with the remaining lines a systematic-uniform distance from the first line. The measured distances are along the lines perpendicular from the vertical axis to the border of the object. For three perpendicular lines, six line lengths will be measured; for four perpendicular lines, eight line lengths will be measured (Figure 7-12).

The gain in efficiency from the rotator relative to the nucleator method stems from the measurement of a larger number of line lengths in relation to the amount of effort required to prepare sections. For each object, with the rotator one obtains six or eight line lengths for the estimate of mean object volume, compared with two line lengths for the nucleator. Thus, with essentially the same effort as measuring two line lengths using the nucleator method, the rotator method captures a greater proportion of the within-cell variation in mean object volume. Repeating this procedure in 100 to 150 objects from the population of biological interest will capture most of the variance in mean object volume. The mean line length is converted to mean object volume using the same formula described earlier for the nucleator estimator.

Like all estimates based on stochastic geometry and probability theory, the unbiasedness of local size estimators is not based on any single volume

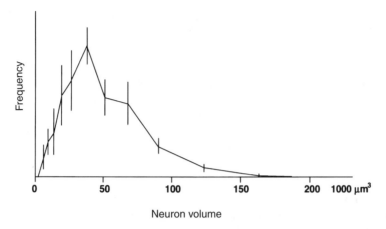

Figure 7-13. Volume distribution for brain cells generated using the rotator-vertical section method

measurement. We do not expect any particular estimate of mean object volume for a particular cell or object to be the correct estimate. Rather, the correct value is the mean object volume for the sample, which if we use a theoretically unbiased method for estimating mean object volume such as the nucleator or the rotator, refers to the expected parameter for the population. For number-weighted estimates of mean object volume, we estimate the mean object volume of a population of objects sampled using an unbiased counting method, for example, the disector principle. With unbiased estimates of total object number and mean object volume, one can generate a volume distribution showing the number of objects across a range of object sizes, as shown in Figure 7-13. The use of either IUR or VUR sections ensures that all possible line lengths through an object's profile will have an equal probability of being included in the sample estimate.

Point-Sampled Intercepts: An Example

In certain situations we are less interested in the mean object volume for an entire population of objects than we are in the mean object volume of the largest object profiles. For instance, in bladder cancer and malignant melanoma, cells with relatively larger nuclei are associated with an increased risk of metastasis. To identify cases that are at higher risk for metastatic disease, it would be useful to know the mean nuclear volume for the largest nuclei in a biopsy sample. While such a probe is not desirable for estimating the mean object volume for a total population of cells, an estimate of mean nuclear volume from biopsy specimens of malignant tumors or precancerous tumors could be helpful to clinicians in the therapeutic management of these types of tumors (for review see Sorensen, 1992).

A single thin, random section through a tissue sample is a volume-weighted probe in the sense that profiles of larger objects are more likely to be hit by a single section than smaller profiles. Similarly, after random placement of a point grid on the surface of a single section, there is a higher probability that points will hit larger object profiles. We can make a volume-weighted estimate of mean object area using the estimator $A = \Sigma P \times a(p)$, in combination with a single thin section and a point grid. In practice, on a tissue section containing object profiles of interest, the reference space is outlined and a point grid is placed at random over the profiles. For each object hit by a point, a line is drawn line from the point to the border in an isotropic direction; that is, all directions to the border have equal probability. In the case of sections through tumor biopsies, if the biopsy is cut at random, then any random direction from the point will be isotropic. Once the first line is drawn and measured, it is efficient to draw a second line in a direction 180° from the first. Both lines will be isotropic, and the mean of the two line lengths will represent the mean line length for the object. This process of measuring line lengths for point-sampled objects is repeated for object profiles through the reference space. The volume-weighted mean object volume, mean v_V, is the product of $\pi/3$ and the individually cubed line lengths (mean l^3) for all object profiles sampled by points. The equation for volume-weighted mean object volume is

$$\text{mean } v_V = \pi/3 \times \text{mean } l^3$$

Thus the larger the objects in the tissue, the greater their probability of being hit by a single 2-D sampling plane, and larger profiles in the tissue will have a greater probability of being hit by a randomly placed point grid. Using a point grid to sample nuclei preferentially samples larger nuclei, thus generating a volume-weighted sample for estimation of the mean nuclear intercepts (Figure 7-14). For many cancers, estimation of mean v_V on thin sections has predictive value for metastatic disease and provides a theoretically unbiased and efficient method for identifying cases likely to require more aggressive treatment.

Ratio Estimators of Object Volume

Point counting provides an efficient method for estimating the area fraction, as described earlier in this chapter ($P_{obj}/P_{ref} = A_{obj}/A_{ref}$). As discussed in Chapter 6, total number can be estimated using the two-stage method ($N = N_V \times V_{ref}$). Similarly, an estimate of total volume can be made with an

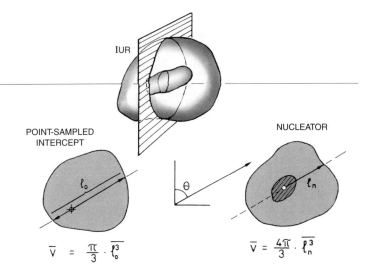

Figure 7-14. Methods of local size estimation on isotropic-uniform-random sections. (From Cruz-Orive and Weibel, 1990.)

estimate of A_{obj}/A_{ref} on single histological sections though a reference space, and a Cavalieri-point counting estimate of the total reference volume, V_{ref}. Recall that $P_{obj}/P_{ref} = A_{obj}/A_{ref} = V_{obj}/V_{ref}$. Thus, using the two-stage formula,

$$V_{obj} = (V_{obj}/V_{ref}) \, V_{ref}$$

where V_{obj} is the total object volume and P_{obj}/P_{ref} is the point fraction determined by point counting.

To avoid the reference trap, both the P_{obj}/P_{ref} and the V_{ref} must be estimated after final histological processing. In this scenario, differential shrinkage or expansion of the objects and the reference space will cancel, thus avoiding the reference trap and providing a theoretically unbiased estimate of total object volume according to the Delesse principle.

Summary

The purpose of using stereological size estimators is to quantify the volume of biologically significant reference spaces and objects from their appearance on 2-D sections without introducing bias from inappropriate assumptions, models, or correction formulas. The use of stochastic geometry and probability theory for sampling and parameter estimation ensures that all parts of the sample will have an equal probability of being included in the sample estimate. These combinations permit investigators to make unbiased estimates of biologically significant volume distributions from systematic-random sections through the reference space in a small sample of individuals. For populations of biological objects (cells, nuclei, etc.), estimators and probes can be combined with volume-weighted sampling (e.g., the point-sampled intersects) or number-weighted sampling using the disector principle.

8

Length and Surface Area

The use of geometric probes to estimate the length or surface area of a biological object in a defined reference space is discussed in this chapter. If a test probe randomly intersects a biological object, the length and surface area of the object are proportional to the number of intersections with the test probe. Geometric probes with two dimensions (planes) and one dimension (lines), respectively, are used to estimate length (L) and surface area (S) on tissue sections. In practice, we calculate local estimators of length density (L_V) and surface density (S_V), then scale these quantities to the total reference volume using either the two-stage method or the fractionator approach.

E*arlier chapters* have described the basic tools of stereology. In this chapter we review the theoretical and practical application of estimators and probes for the two remaining first-order stereological parameters: length and surface area.

For insight into modern stereological approaches for theoretically unbiased estimates of length and surface area, it is useful to review the principles of stochastic geometry and probability theory in Buffon's needle problem (Chapter 4), which asks: *What is the chance that a needle randomly tossed into the air will fall across the lines on a parquet floor?*

Buffon's solution is: *The needle will intersect lines with a probability that is directly proportional to the length of the needle and inversely proportional to the distance between the lines on the floor.* In mathematical terms, the Buffon principle can be restated as follows:

$$P = (2/\pi) \times (l/d)$$

where *P* is the probability of a needle–line intersection, *l* is the length of the needle (cm), and *d* is the distance between the lines on the floor (cm).

For the number of needle–line intersections to be proportional to length, the parameter of interest, all intersections must occur with a random probability. Random probability for needle–line intersections is ensured by the random toss of the needle into the air. The factor $2/\pi$ in the Buffon principle integrates all the possible angles of intersection between the needle and the lines on the floor in 2-D, that is, from $0°$ to $360°$. Provided the needle hits *somewhere* on the parquet floor, the probability of an intersection is proportional to the length of the needle and the distance between the lines on the floor. By actually tossing the needle a minimum number of times (e.g., 100 to 200 times), we can count the number of intersections observed between the needle and the lines on the floor. From these data we can calculate *P,* the probability of an intersection, as the ratio of intersections to tosses. Once this probability is known, if one also knows either the true length of the needle or the true distance between lines on the floor, one can solve the equation for the missing quantity in a theoretically unbiased manner, as shown in the following example:

> Consider an object of unknown length, *l,* which is probed with test lines separated by 10 cm ($d = 10.0$ cm). For a theoretically unbiased estimate of *l,* we repeatedly toss the needle in the air and observe the number of intersections with lines on the floor. The sum of needle–line intersections (ΣI) divided by the sum of tosses (ΣN) gives us a ratio ($\Sigma I/\Sigma N$), the probability, *P,* of an intersection. For instance, 25 intersections out of a possible 100 tosses gives an intersection probability of $25/100 = 0.25$. Finally, the distance, *d,* between the lines in the line probe is constant at 10.0 cm. Entering *P* and *d,* we can solve for *l:*
>
> $l = (\pi/2) \times (I/N) \times d$
>
> $l = 1.6 \times 0.25 \times 10.0$ cm
>
> $l = 4.0$ cm

Of course Buffon's needle problem did not provide the first or even the best method available for measuring length. Two thousand years before Buffon presented his problem in Paris, Euclid produced his book *Elements,* which contained a variety of model-based methods for quantifying the length of rectangles and squares, the perimeter of circles, and the length of the hypotenuse of a triangle. What Buffon's principle does do for the first time, however, is to allow the known length of a nonclassically shaped object to be estimated based on the laws of probability, without any assumptions about the shape of the object. Although a straight needle is used in the needle problem, the method is accurate for estimating the length of any population of objects, regardless of their shape. The objects of interest can be curved at any angle or direction. Furthermore, whether three objects are thrown a hundred times or a hundred objects are thrown three times, provided the throws are random, the number of probe–object intersections will be proportional to the total length of the object(s), without further assumption. Thus, Buffon's needle problem provides the theoretical basis for estimating the length or surface area (boundary length) of populations of nonclassically shaped objects of biological interest, based on their appearance on 2-D tissue sections.

The estimation of boundary length (i.e., the object's surface area) follows directly from the previous discussion of object length. Recall the solution to the needle problem:

$$l = (\pi/2) \times (I/N) \times d$$

Imagine that N tosses of the needle, when placed end to end, create a randomly shaped curve on the floor, as shown in Figure 8-1. Taking the total length of the needles as the product of the number of tosses, N, and the length of one needle, l, we can substitute $B = N \times l$, and solve for the total boundary length, B,

$$B = (\pi/2) \, I \times d$$

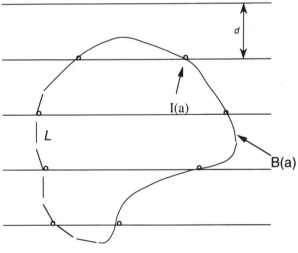

$$L = (\pi/2)(I/N) \, d$$

$$\text{If } B = N \, L, \text{ then}$$

$$B = (\pi/2) I d$$

Figure 8-1. Unbiased estimation of surface area (boundary length, B) is based on a probability orginally suggested by Buffon's needle problem. (From Weibel, 1979.)

Now we have a theoretically unbiased method for estimating the boundary length (surface area) of a curvilinear profile. Because the linear features and surface areas of biological objects appear on tissue sections as curvilinear profiles, this approach provides the basis for theoretically unbiased estimates of surface area (S) and length (L).

From 2-D to 3-D

Note that in contrast to biological tissue, in the needle problem the probability of intersections takes place entirely in two dimensions. Therefore, the sum of dimensions in the probe (dim = 1) and the parameter (dim = 1) must be equal to 2. To understand how stochastic geometry and probability theory inherent to the needle problem apply to biological objects, we must convert our thinking from 2-D to 3-D. For biological tissue in which the parameters of biological interest are defined in 3-D space, which includes all biological objects, the sum of probe + parameter dimensions must equal or exceed 3.

Suppose we are interested in the total length, L, of a population of thin, curvilinear objects in a defined reference space. In place of a needle, we can substitute "population of biologically interesting curvilinear objects" in a restatement of the needle problem: *What is the probability that a population of biologically interesting linear objects randomly tossed into the air will fall across the lines on a parquet floor?*

Instead of using the lines on a parquet floor to estimate L, let us substitute "parallel lines on a grid" in reference to the parallel lines in a line grid that are separated by a known distance, d: *What is the chance that a population of biologically interesting linear objects randomly tossed into the air will intersect parallel lines on a grid separated by distance d?*

One can appreciate that we are deriving a stereological estimator for total length. First, however, the formula must take into account the fact that the linear features and surface areas of biological objects occur in 3-D, rather than the 2-D planar surface of a parquet floor.

Unbiased Stereology Requires 3-D Thinking

The key to applying Buffon's principle to biological tissue lies in the random toss, which in stereological terms refers to isotropy in 3-D. A process or object is isotropic when all orientations have an equal probability in 3-D space. Thus the starting point for applying Buffon's principle to biological tissue is ensuring isotropic intersections between the probe and the object of interest.

All the action in the needle problem takes place in 2-D. Both the probe (grid lines) and the linear object (needle) are 1-D; thus the sum of dimensions in the probe and the parameter equals 2. In contrast, biological objects evolve in the 3-D confines of tissue. However, the process of identifying microscopic biological objects encourages a 2-D perception. The visualization of microscopic biological objects requires that the reference space be divided into thin sections that facilitate the passage of images (via light, laser, electrons) to the eye for identification. This process causes a distortion in which 3-D objects appear flat. Estimators and probes that fail to take the full dimensionality of 3-D space into account contain an unknown quantity of stereological bias.

When lines are moved in space, they become 2-D planes. Therefore a theoretically unbiased geometric probe for estimating length in 3-D can be a grid of parallel planes formed by moving lines in space. Substituting this statement into the original statement of the needle problem: *What is the chance that a population of biologically interesting linear objects randomly tossed into the air will intersect 2-D planes?*

Isotropic Probe–Object Intersections

For theoretically unbiased biological applications of the needle problem, probe–object intersections must be isotropic, meaning that intersections in all directions in 3-D must have an equal probability of occurring. This requirement ensures that the shape and 3-D orientation of the object have no impact on the probability of an intersection. As for all theoretically unbiased stereological probes, only three factors can influence the number of probe–object intersections: (1) the magnitude of the parameter, (2) constants required to convert the number of intersections into the required parameter units, and (3) known dimensions of the geometric probe.

The needle problem has the same requirement, but only in the two dimensions where possible intersections can occur. All possible angles of needle–line intersections are ensured by randomly tossing the needle into the air. For estimating the length of curvilinear objects, isotropic intersections must occur in 3-D. To complete the conversion for application to populations of biological objects, we can incorporation this requirement into the needle problem: *What is the probability that a population of biologically interesting linear objects will make isotropic intersections with 2-D planes?*

Optimal Levels of Sampling

From this discussion of probability we can gain insight into the basis for the adage to do more less well; that is, to capture most of the variation in a parameter using a minimal amount of time, effort, and resources. The starting point for optimization of stereological designs is setting the dimensions of the geometric probe so that about 100 "hits"—probe–object intersections— occur. A pilot study in which 100 hits are counted is analogous to estimating the length of a needle based on 100 needle–line intersections. The number of tosses needed to achieve 100 hits depends on the distance between the lines on the floor; the closer the lines, the smaller the value of *d*, and the greater the probability of a needle–line intersection. Finding this optimal distance for the line probe requires a bit of trial and error at first, which is analogous to a pilot study. Once this optimal distance is determined, the parameters are set for the analysis of linear objects of a variety of lengths.

One might wonder whether a better estimate would result from increasing the number of hits from 100 to 200, 300, 500 or more. Recall that the number of observed probe–object intersections is used to calculate the probability of an intersection. From this probability and one or more constants, the parameter of interest is estimated. Therefore we might rephrase the question as, "Would I expect to obtain a better estimate of the probability of an intersection if I tossed the needle 500 times to calculate the probability of an intersection with the parquet floor?" For the sake of discussion, consider the data in Table 8-1.

As shown by these hypothetical results, after about 300 tosses, the probability of an intersection converges on a stable value of 20%. One could make the argument that more than 150 tosses are required because the number of hits observed at this low intensity of sampling does not result in a stable estimate of *P*. Beyond 200 or so, the percentage of tosses resulting in an intersection, and therefore the probability of an intersection, remains stable. With a stable estimate of *P*, from Buffon's needle we can make a theoretically unbiased estimate of *l*, the length of the needle.

Table 8–1 Probability of Intersection across a Range of Needle Tosses onto a Floor with Constant Distance between Lines

No. of tosses	No. of hits	Probability (*P*)
50	14	14/50 = 0.28
100	25	25/100 = 0.25
150	33	33/150 = 0.22
200	44	42/200 = 0.21
300	60	60/300 = 0.20
500	100	100/500 = 0.20
750	150	150/750 = 0.20
1000	200	200/1000 = 0.20

Because the parameter of biological interest is unknown at the start of a study, the optimal probe dimensions and spacing between probes to achieve 100 hits must be determined by trial and error. Finally, because geometric probes are placed over a defined area of reference space, knowing the optimal number of probes required to achieve 100 hits allows the optimal probe size and spacing between adjacent probes to be determined. Therefore, increasing the size of the probe or reducing the spacing between probes will result in a greater number of hits. However, as illustrated by the above discussion and the following empirical data, the added counts do not appreciably change the probability of an intersection and therefore have little effect on the parameter of interest. Thus, the essential purpose of a pilot study is to determine the optimal dimensions of the probe and the spacing between adjacent probes to achieve about 100 hits and a stable parameter estimate. Beyond this optimal level of sampling, the rational strategy is to discontinue sampling in the same individual; in stereological terms, the optimal level of sampling has been achieved.

As illustrated by Figure 8-2, counting up to about 200 cells leads to a stable estimate of total cell number in a defined reference space. It is important to note that counting more than 200 cells does not appreciably change the estimate of total cell number. As discussed in Chapter 3, Gundersen and Jensen showed that the optimal level of systematic-random sampling in the reference space of any individual, regardless of the parameter or any particular

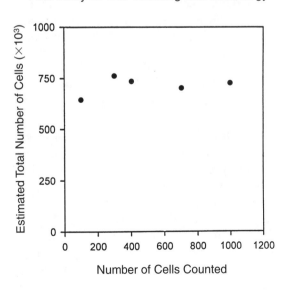

Pilot Study for Cell Counting with Stereology

Figure 8-2. Actual data showing that counting a maximum of 200 cells leads to a stable estimate of total cell number for a defined brain region. (Figure supplied courtesy of Dr. Kebreten Manaye, Howard University. Reprinted with permission.)

morphological features or spatial distribution of objects, is less than 200 probe–object intersections. Beyond this point, the rational strategy is to efficiently reduce the variation observed in the sample estimate by sampling lightly in more individuals from the population.

Orientation Bias in Surface Area and Length

The requirement of isotropy for probe–object intersections applies only to estimation of length and surface area parameters; estimates of volume (V) and number (N) are orientation independent. We can understand the reason for this mathematically as well as intuitively; for simplicity we consider the latter explanation.

Imagine a series of linear objects oriented parallel to each other in tissue. To quantify the total length of the linear objects, we must first cut the tissue into sections, place a line probe over the cut surface, and count the number of probe–object intersections. However, the probability of probe–object intersections will differ, depending on whether the tissue is sectioned parallel or perpendicular to the long axis of the objects. The same situation applies to the surface area of a biological object. The length and surface area of objects occupy one or two dimensions of the possible three dimensions in tissue, respectively, leaving two (length) or one (surface area) unaccounted-for dimension(s) in three-dimensional space.

The factor that determines whether probe–object intersections occur in these unaccounted-for dimensions is not the magnitude of the parameter, as required for a theoretically unbiased method; instead, the number of probe–object intersections is determined by the orientation of the tissue sectioning. A number of approaches have been developed in recent years to overcome this potential bias arising from the orientation of tissue sectioning. Before reviewing these procedures, however, it is useful to reiterate that this bias related to direction of sectioning applies only to the parameters of length and surface area; number and volume are not affected by the direction in which tissue is sampled.

Again, we can understand this distinction intuitively. While for length and surface area the probability of probe–object intersections is a function of the direction of tissue sectioning, for volume, the probability of probe–object intersections is directly related to the volume of the object in the tissue. The greater the object's volume, the greater the probability that the object will appear on the section, regardless of the direction of cutting. There is no bias related to sectioning because the total dimensions in the probe, the 2-D plane of the knife, and the parameter volume (3-D) sum to more than 3. As we section through tissue containing an object, the entire volume of that object will be sampled. Similarly, as we section through tissue containing a number of objects, the entire number of objects in the tissue will be revealed. The parameters of volume and number apply to 3-D objects, whereas surface area and length apply to two- and one-dimensional objects, respectively.

Because surface (2-D) and length (1-D) refer to objects in a section that occupy fewer than three dimensions, the direction from which linear objects and surfaces are sectioned has a significant impact on the probability of intersections and hence on the magnitude of the estimate. Thus, we say that the surface area and length of biological objects are anisotropic; their spatial orientations in 3-D have unequal probabilities. To ensure isotropic probe–object intersections as required by the Buffon principle, stereologists have devised a number of ingenious methods to overcome the inherent anisotropy of biological objects.

Overcoming Anisotropy of Biological Tissue

When anisotropic surfaces and linear objects in tissue are cut into sections, their inherent anisotropy is transferred to 2-D sections. Thus special procedures are needed to avoid the introduction of stereological bias. This can be done using two approaches: randomize the orientation of the object or randomize the orientation of the probe. Randomizing the orientation of both the probe and the objects is unnecessary and adds no additional benefit; to overcome anisotropy, either the object or the probe must be isotropic in 3-D to ensure isotropic probe–object intersections. In the following section we review older approaches to overcoming anisotropic surfaces and linear objects by randomizing the orientation of tissue; then we outline newer strategies that involve the use of isotropic probes.

Randomization of Objects

The first method for overcoming anisotropy involves randomization of the objects in tissue prior to sectioning by making isotropic-uniform-random sections (see Chapter 7). On IUR sections, isotropic probe–object intersections occur when a straight-line test probe is superimposed on objects on the sections. Like the random toss of a needle, IUR sections ensure that all orientations of the object have the same probability of intersecting the line probe. The problem for many biological applications, however, is that disorientation of tissue landmarks on IUR sections renders recognition of objects of interest difficult, tedious, and sometimes not possible, particularly in the case of layered structures. To minimize the effect of this disorientation on anatomical structures, a second method for overcoming tissue anisotropy uses a special line probe called a cycloid. In 1986 Baddeley and co-workers proposed using this probe with vertical-uniform-random sections to estimate surface area. In 1990 Gokhale showed that cycloids in combination with VUR slices provided a theoretically unbiased approach for estimating length.

IUR Sections and Straight-Line Probes

The isotropy in IUR refers to the orientation of biological objects in tissue with regard to the section plane in which they are cut. The advantage of an IUR section is that the orientation of objects within the tissue is random with regard to the cutting direction. The simplest strategy is to completely randomize the tissue in space prior to sectioning. IUR sections can be difficult to make, however. One must take care to avoid preselection of orientation (e.g., coronal, horizontal, semihorizontal), which might preserve the natural anisotropy of biological surfaces and linear objects on tissue sections and introduce stereological bias into surface area and length estimates.

A convenient method for creating IUR sections is called the isector (Nyengård and Gundersen, 1992). This involves embedding tissue containing the object of interest in preformed spherical molds containing unhardened paraffin or a plastic embedding medium. After the medium hardens, the tissue is isotropically embedded in a spherical block, which can be randomized in 3-D (rolled) before sectioning. Sectioning is systematic-random through the entire reference space. For IUR sections, the test probe is placed over the objects at random and the random probe–object intersections are counted.

Thin IUR Sections as Unbiased Sampling Planes for Length Density

Since length is a 1-D parameter, its estimation requires a 2-D sampling probe (plane) to be theoretically unbiased. The length density, L_V, refers to the length of a linear object per unit of tissue volume (V). The surface of a tissue section is a useful sampling plane for estimation of L_V; the planar surface of sections will hit 1-D profiles of linear objects with a probability that is proportional only to the total length of the objects. Note that in this approach we are estimating the length density of linear objects of interest based on the number of profile intersections within a sampling plane of known area. In stereological terminology, the term Q refers to the number of intersections between an object's profile and the planar surface of a sampling plane. The designation Q^- refers to new intersections between a profile and the sampling plane; new means that profiles from the same object are not present on the previous observation plane.

It is unnecessary to count all the profiles on the surface of sampling planes through a reference space; therefore, profiles are counted on a sample of the reference area. Profiles are counted for the first time only once to avoid stereological bias arising from the size, shape, and orientation of the profile. Counting the number of object profiles in a theoretically unbiased manner requires unbiased counting rules and an unbiased counting frame, as discussed later.

Finally, profiles are counted on about 100 to 150 unbiased counting frames spaced in a systematic-random manner throughout the entire reference space. As detailed in Chapter 10, counting between 100 and 200 intersections is sufficient to capture the within-sample variability for essentially any parameter of biological importance. Thus, counting between 100 and 200 unbiased counting frames spaced in a systematic-random pattern throughout the reference space is the optimal sampling intensity for making theoretically unbiased estimates of the L_V in a defined reference space.

A couple of observations deserve clarification. First, because the relationship between the length of a linear structure and one profile is 1 to 1 (one profile = one length of the object), the total number of profiles hit by the sampling plane is proportional to the total length of the linear object. Second, the probability of hitting the area of a reference space is proportional to the volume of the reference space, $A_A = V_V$, according to the Delesse principle. Therefore, although we are actually counting the number of object profiles in a known reference area, Q_A, this ratio is proportional to the length of the object profile per unit of reference space volume (Smith and Guttman, 1953).

Finally we have arrived at the critical formula for estimation of length density (μm^{-2}, mm^{-2}, cm^{-2}) for a linear biological structure. This formula shows that the number of linear profiles per unit of the section area, Q_A, is directly proportional to the length density, L_V, of linear objects per unit of tissue volume. The formula for estimation of L_V is

$$L_V = 2\,Q_A = 2 \times \Sigma Q^- / \Sigma A$$

where ΣQ^- is the sum of the number of length–plane intersections through the reference space and ΣA is the sum of the area of the counting frames counted (μm^2, mm^2, cm^2).

Point counting is a convenient method for estimating ΣA, the area of the reference space sampled. Rather than counting sampling frames, one simply uses a point grid of known area per point and counts one point within each sampling frame. In this way we can keep track of the sum of points, ΣP, hitting the reference area, along with the number of profiles intersected by the sampling plane. The quantity ΣA is the product of the number of points counted and the area per point: $\Sigma A = \Sigma P \times$ area per point.

$$L_V = 2Q_A = 2 \times [\Sigma Q^- / (\Sigma P \times \text{area per point})]$$

Thus, estimation of L_V by using a section plane to sample linear profiles simplifies to two observable events—the total number of probe–object intersections and the total number of points hitting the reference area where the objects are sampled.

When using a 2-D plane to estimate the length of linear profiles within a counting frame, we use counting rules to ensure that all profiles have the same probability of being sampled and to avoid multiple counts at the edge of the counting frame, the edge effect. The unbiased counting frame and the unbiased counting rules were specifically developed by Gundersen in 1977 to avoid these potential sources of stereological bias when counting object profiles. As detailed in Chapter 6, profiles that fall within the counting frame or touch the inclusion lines of the sampling frame are counted, while profiles that touch the exclusion lines of the counting frame are ignored.

Vertical Sections and Cycloids

Two significant problems have limited wider use of IUR sections for estimation of surface area and length. The first problem is related to the disorientation mentioned earlier; biologists prefer to observe morphology on sections in a fixed, preselected orientation (e.g., coronal). The difficulty imposed in creating true IUR sections has frequently led scientists to avoid estimating surface area and length, or worse, encouraged them to make assumption- or model-based estimates without regard to stereological bias. A second disadvantage is that although the approach ensures unbiased estimates of surface area and length, a substantial amount of sampling is typically needed for precise sample estimates with IUR sections. These sections are relatively inefficient when making sample estimates because only a single orientation can be contained within a given section. Capturing a greater percentage of the inherent biological variability requires a relatively large number of cases.

In VUR sections, however, one axis is preselected as "vertical." This axis becomes the focal point around which the tissue containing the objects of interest is rotated (rolled) at random prior to being sectioned. For optimal efficiency, the selection of the vertical axis should take into account the preferred direction of the objects of interest. If the structure is composed of many objects with somewhat variable orientations, the most efficient vertical axis is the axis shared by the majority of the objects.

Imagine a sample of leaves on trees blowing in a strong horizontal wind. Although their orientations will vary from 0° to 90° relative to the wind direction, the predominant orientation of the leaves will be parallel to the direction of the incoming wind. Note that "vertical" in this terminology bears no relationship to vertical in relation to gravity or up and down. It is simply the axis selected by the investigator around which the tissue is randomly rotated prior to sectioning. This rotation effectively randomizes the objects' orientation around the vertical axis. When selecting a vertical axis, the most efficient choice will be the axis that contains the least amount of variability in the parameter of interest. In the example of the leaves, that axis is the predominant long axis of the leaves, that is, parallel to the blowing wind. If the wind were not blowing, then the best choice for a vertical axis might be true vertical relative to gravity. Once rotation is complete around the vertical axis, the tissue is then sectioned in a direction that is perpendicular to the vertical axis. The result is that the tissue sections will contain profiles of objects cut parallel to the vertical axis, that is, where they show the least variability.

In addition to a preselected axis, VUR sections also use a special line grid called a cycloid to make sample estimates of surface area and length. A cycloid is a classical geometric line that has a special property—it curves according to a sine-weighted orientation. The sine-weighted lines compensate for the preselected, nonrandom vertical axis of a VUR section (Figure 8-3). Instead of a smooth line curving from vertical at 0 to 90°, a cycloid has a short initial length where the angle is near 0° (i.e., parallel to vertical), and the segment progressively lengthens as the angle from vertical approaches 90°. A cycloid can be thought of as a group of small line segments lined up end to end, with each new segment having an angle increasing from vertical of 0 to 90°. The greater the angle from vertical, the longer that segment of the cycloid, which increases proportionally. The longest part of a cycloid is where the line segment is perpendicular to the vertical axis. Mathematically speaking, each length of segment within a cycloid is proportional to sin θ, where θ is the angle of that line segment from the vertical axis (Figure 8-4).

As shown in Table 8-2, the length of the cycloid increases nonlinearly according to the sine of the angle θ relative to the vertical axis. Imagine standing in the center of the Earth and rotating yourself in a circle. At some point

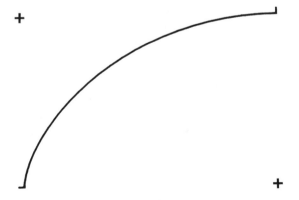

Figure 8-3. A cycloid is a sine-weighted line

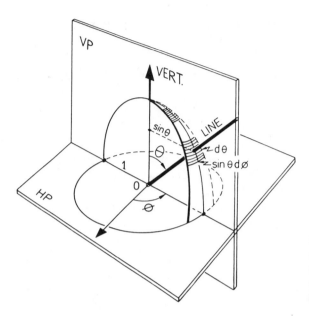

Figure 8-4. Components of a cycloid in the vertical plane (VP) and horizontal plane (HP). (From Baddeley et al., 1986.)

you stop and extend your arm in a random direction. The tip of the finger on your extended hand points toward a random spot on the Earth's surface. Every location on the Earth's surface is defined by its x,y,z coordinates in space. However, you would not point at all x,y,z coordinates on the Earth's surface with an equal probability. Instead, there will be a higher chance of pointing to areas near the equator, and a lower chance of indicating points near the extreme North and South poles. The purpose of a sine-weighted test line (cycloid) is to transfer this differential sampling probability to a 1-D line. In conjunction with VUR sections, a cycloid ensures random probe–object intersections for the estimation of surface area and length. Using the approach introduced by Baddeley for surface estimation and used by Gokhale for length estimation, the combination of cycloids and VUR slices or sections allows experimenters to preselect at least one axis in the tissue.

Table 8–2 Progression of θ and Sine θ from 0 to 90°

Angle θ relative to the vertical axis	sine θ
0	0.00
22.5	0.38
45	0.71
67.5	0.92
90	1.00

Thin Focal Plane Scanning for L_V

A grid of 1-D line probes does not provide a theoretically unbiased estimate of the length of 1-D linear objects; the sum of probe–parameter dimensions is only 2. In 1990 Arun Gokhale published a method to estimate L_V from projected images of linear objects using cycloids and VUR slices. The Gokhale approach takes advantage of the 2-D projection of a cycloid line into a "cycloid plane" to make theoretically unbiased estimates of L_V. Using a 1-D cycloid test line focused through the z-axis of VUR slices, the area traced by the test lines resembles a 2-D plane. That is, in a defined reference space, the number of intersections between a 1-D linear object and the 2-D surface of a cycloid plane is proportional to L_V.

The formula for estimating L_V using sampling planes moved through the z-axis of thick sections is

$$L_V = 2 \times 1/h \times \Sigma I / \Sigma P$$

where ΣI is the sum of the length–plane intersections on all sections, ΣP is the sum of points hitting the reference area (μm^2, mm^2, cm^2), and h is the height sampled in the z-axis (z-height).

Thin focal plane scanning is analogous to scanning using the 3-D optical disector to estimate the numerical density of objects, N_V. Instead of counting the tops of objects intersected by the virtual scanning plane in a known volume, Gokhale's method counts the number of sampling plane–linear object intersections in a known volume. Recall that density estimators such as N_V and L_V are ratios; like all ratios, density estimators are subject to changes in both the numerator and the denominator. That is, L_V and N_V provide no independent information on total length or total number without assumptions about the denominator, the reference volume. As seen earlier, the assumption that reference spaces do not change, or that they change consistently in different tissues, leads to potential bias (the reference trap). To avoid the inherent bias of ratio estimators, L_V is converted to total length, L. In the following sections we describe two approaches for calculating total length. As

Table 8–3 Scaling of Ratio Estimators to Absolute Parameters

Parameter		Two-Level Design
Total N	=	$(N_{obj} / V_{ref}) \times V_{ref}$
Total S	=	$(S_{obj} / V_{ref}) \times V_{ref}$
Total L	=	$(L_{obj} / V_{ref}) \times V_{ref}$
Total V	=	$(V_{obj} / V_{ref}) \times V_{ref}$

shown in Table 8-3, the same approaches can be used to scale ratio estimators involving other parameters to the desired first-order stereological parameters.

Two-Stage Calculation of Total Length from Length Density

Total length, L, is the product of the length density (L_V) and the total volume of a defined reference space (V_{ref}):

$$L = L_V \times V_{ref}$$

Total reference volume can easily be estimated by point counting and the Cavalieri principle (see Chapter 7). Because both V_{ref} and L_V refer to the same volume containing the linear objects of interest, their product cancels any changes (shrinkage or expansion) that occur during tissue processing, leading to a theoretically unbiased estimate of total length. For this cancellation to be valid, however, both L_V and V_{ref} must be measured on the same tissue following identical histological processing.

Estimation of Total Length using the Fractionator Principle

The fractionator approach is an alternative method to the two-stage L_V $\times$ V_{ref} approach for scaling local parameter estimates to the reference space. If L_V is estimated in a known fraction of the total reference volume, then multiplication of the local estimate by this fraction will provide a theoretically unbiased estimate of total length. According to the fractionator approach, the reference volume containing the linear object of interest is first serially sectioned in its entirety. A known fraction of the total sections (e.g., one-tenth) is sampled in a systematic-uniform-random manner. This sampling interval is the section sampling fraction. Typically the interval is selected to obtain between 8 and 12 sections through the reference volume. The first section in the first interval is taken at random; then sections are taken at the same sampling interval throughout the reference space. For instance, if the reference volume is contained in a total of 600 sections, to sample a theoretically unbiased set of 10 sections, the sampling interval will be every sixtieth section. In that case the *ssf* will be 10/600 = 0.0167. To ensure that the sampling is theoretically unbiased, the first section in the first interval of sections 1 to 60 is taken at random (e.g., section 24); from this point, sections are taken every 60 sections through the total number of sections containing the reference volume. If the length of the linear objects in these sampled sections were known, then the product of this quantity and the reciprocal of the *ssf* (i.e., 1/*ssf*) would provide a theoretically unbiased estimate of the total length of the linear objects in the reference volume.

However, it is not efficient to quantify the total length for the entire reference area on each of 10 sections sampled through the reference space. A theoretically unbiased estimate can be made by sampling a known fraction of the total reference area on each section, the area sampling fraction (*asf*). This refers to a systematic-uniform-random sample of the reference area on each section. Like the *ssf*, the *asf* is a ratio—the area sample over the total reference area—and has no units. A sampling frame of known area (frame area = *x* length of frame × *y* length of frame) is moved in a systematic-random manner through known x and y distances in the tissue (area step = distance in the x-direction × distance in the y-direction). This process is repeated across the total reference areas on the sections from the *ssf*. At each x-y location, the sampling frame is placed over the reference area. Finally, we obtain a theoretically unbiased estimate of total length for the individual by scaling this local estimate to the total reference volume. Thus, using the fractionator method in conjunction with Gokhale's estimator for length using VUR slices and cycloids provides theoretically unbiased estimates of total length from the sum of probe–object intersections within a known fraction of the reference space.

Using the fractionator approach described above, total length is proportional to the product of ΣI and the reciprocals of the two sampling fractions, $1/ssf \times 1/asf$. When total length is estimated using the thin focal plane scanning method on thick sections, a third sampling fraction is required. This fraction, the thickness sampling fraction (*tsf*) is the distance scanned in the z-axis (i.e., the height, *h*, of the disector) divided by the total section thickness, *t*. This fraction (h/t) is the region where probe–object intersections (*I*) are counted to obtain the local estimate. Finally, the sum of intersections (ΣI) counted in this local estimate is scaled to total *L* by multiplying ΣI and the reciprocals of all three sampling fractions (*ssf, asf, tsf*). The formula for estimating total length using the fractionator method in conjunction with thin focal plane scanning, VUR slices, and cycloids is

$$L = 2 \times \Sigma I \times a/l \times F_1 \times F_2 \times F_3$$

where:

F_1 = 1/section sampling fraction = 1/(number sections sampled/ total sections)

F_2 = 1/area sampling fraction = 1/(area of section sampled/total area)

F_3 = 1/thickness sampling fraction = $1/(h/t)$

(a/l) = area per test line on test grid (μm, mm, cm, etc.)

The term (a/l) is the grid constant that relates the length of the cycloid test line to the area on the test grid. This term is analogous to the constant distance between the line on the floor in Buffon's needle problem. When the linear objects are particularly dense, proportionally less test line per unit area is needed than when the linear objects are less dense. Changing the grid constant allows changes in the amount of test line on the grid without introducing bias into the estimate.

For the fractionator approach to give theoretically unbiased estimates, it is necessary to know the *ssf*, the fraction of sections sampled in the total number of sections containing the reference space. Thus the approach is not appropriate for exceedingly large structures in which serial sectioning is impractical. When serial sectioning is practical, application of the fractionator approach to either thin or thick sections provides the most efficient method for scaling local estimates to the reference space.

Estimation of Surface Area

Surface area refers to the total surface area of an object or a population of objects contained within the reference space. Because it is a 2-D parameter, theoretically unbiased estimation requires surface probes that are at least 1-D. Otherwise the approaches for estimating total surface area are analogous to that described earlier for estimating total length.

Imagine tossing test lines of known length into a defined space containing objects with surfaces of biological interest. Provided the length of the test lines remains constant, the probability of isotropic intersections between the object's surface area and the test lines will be proportional to the total surface area of the objects, with no further assumptions (Figure 8-5).

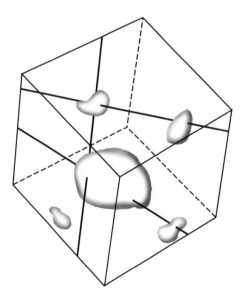

Figure 8-5. Lines placed at random intersect the surface of objects with a probability that is proportional only to the total surface area of the objects, with no further assumptions.

In practice, the tissue containing the surface of interest is first sliced into thick slabs or thin sections. A grid of test lines is oriented at random on the cut surface of the sections and intersections are counted when test lines cross the surface of the objects. On 2-D slices, surfaces appear as lines surrounding a contained volume of tissue. The number of random intersections between a probe (test line) and an object (surface area) will be directly proportional to the surface area of the object.

The sum of intersections per unit of test-line length is directly proportional to the surface density of the object (S_V) through the total reference space. S_V is equal to the sum of surface area–test-line intersections (ΣI) per unit of test-line length (ΣL), according to the formula:

$$S_V = 2 \times \Sigma\, I / \Sigma\, L$$

where $\Sigma\, L$ is the total length of test line summed over all sections (μm, mm, cm).

In practice, it is convenient to use a test-line grid in which a point (+) is associated with each test line (e.g., "------+"). In this way the investigator can simultaneously record the total length of the test line (ΣL) while counting probe–surface intersections (ΣI); the total length of the test line is the product of the number of points counted (ΣP) and the length of one test line (l); $\Sigma L = \Sigma P \times l$. The length of one test line refers to the actual length in the tissue, that is, after correction for magnification. Thus, the process of estimating the S_V of objects on tissue sections reduces to two observable events: the number of test-line–surface intersections on all the sections (ΣI) and the number of points that hit the volume of the object (ΣP). Finally, the total surface area (S) is calculated as the product of surface density (S_V) and total reference volume (V_{ref}):

$$S = S_V \times V_{ref}$$

For total surface area, the equation for the fractionator-based approach is the following:

$$S = 2 \times \Sigma I \times a/p \times F_1 \times F_2 \times F_3$$

where F_1, F_2, and F_3 are as defined earlier and (a/p) is the area per point (μm^2, mm^2, cm^2) on the test grid.

As mentioned earlier in this chapter with regard to total length, for theoretically unbiased estimates of surface area, probe–object intersections must be isotropic because of the 3-D anisotropy that is transferred to tissue sections in 2-D.

Test-Line Orientation

In estimating surface area and total length using IUR sections, the tissue is probed using a grid of straight lines analogous to the lines on a parquet floor. Recall that because IUR sections randomize the orientation of objects, the orientation of the test lines used to probe the objects is not critical. Grids of straight lines are simply placed over the IUR sections at random and the probe–object intersections are counted. For VUR sections, however, the orientation of objects in the tissue is not 100% randomized; thus the cycloid must be properly oriented relative to the vertical axis to complete the randomization process. The proper alignment of cycloid grids for estimation of surface area and length requires that the cycloid be divided into its component parts.

The major axis of the cycloid is the long side (length); the minor axis is the short side (height). In estimating S_V using vertical sections, the *minor* cycloid axis is oriented parallel to the vertical axis (Figure 8-6). In estimating L_V the *major* cycloid axis is oriented parallel to the vertical axis (Figure 8-7). In practice, the investigator superimposes the line grid over an image containing the object of interest and counts the number of line–object intersections.

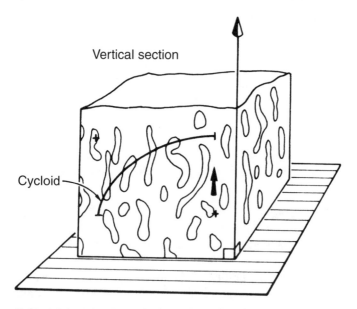

Vertical section

Cycloid

Figure 8-6. In estimating surface area, the long (major) cycloid axis is oriented perpendicular to the vertical axis (arrow) using the VUR section/cycloid approach. (From Cruz-Orive and Weibel, 1990.)

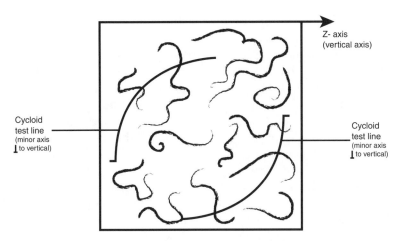

Figure 8-7. In estimating length, the short (minor) cycloid axis is oriented perpendicular to the vertical axis (arrow) using the VUR slice/cycloid approach. (From Gokhale, 1990.)

Grids can be manually placed over photomicrographs, taped over a digital image on a computer screen, or automatically oriented over digitized images using software programs. As usual for all probe–object intersections, the cycloid grid is placed over the objects at random, that is, without any attempt to influence intersections. For noncomputerized approaches, the investigator simply places the line grid over the section without regard to the objects in the tissue. Although it is important to ignore objects in the tissue when placing grids for counting intersections, as explained earlier for VUR sections, it is important that the investigator carefully orient cycloid grids parallel to the vertical axis.

Thus, the VUR section + cycloid combination provides an efficient alternative to the IUR section + straight test-line approach. Both approaches effectively randomize the intersection between the geometric probes (line) and the surface area or linear feature of interest. The estimates of surface area and total length using these approaches are theoretically unbiased, so as sampling through the tissue increases, the sample estimates will progressively converge on the true or expected parameter for the population.

The methods discussed thus far for ensuring isotropic probe–object intersections, as required for theoretically unbiased estimation of stereological parameters, involve randomization of tissue around all three x,y,z-axes (IUR sections) or one vertical axis (VUR slices or sections). In the past 3 years, several novel approaches have been developed for rendering isotropic probes; these are discussed in the remainder of this chapter.

Estimating Surface Area and Length Using Isotropic Probes

For all unbiased approaches to estimating first-order parameters of biological interest, the goal is to estimate the parameter with a probe that intersects the object of interest with a probability that is proportional to the true magnitude of the parameter. For estimating the first-order parameters of number and volume, this goal will be met regardless of the geometry between the object of interest and the scanning probe. That is, regardless of the angle at which we approach a group of objects with a 3-D disector probe, the number of object–probe intersections will be directly proportional to the true number of objects, with no further qualifications. Similarly, we can approach a volume of tissue on sections with a point grid oriented at any angle, and the number of points that hit the tissue will be directly proportional to the volume of tissue. Thus, the parameters of number and volume are orientation independent.

As mentioned previously, the same situation does not apply to the first-order stereological parameters of surface area and length. Planar surfaces and linear features of biological origin are more or less orientation dependent. We must first recognize the problem that planar surfaces and linear features of biological origin are inherently anisotropic; that is, they are not equally oriented in all directions, but rather show a preferred orientation in 3-D space. If we approach a 1-D linear feature in space with a 2-D planar probe, or a 2-D planar surface with a 1-D line probe, the anisotropic orientation of the biological features will strongly influence the probability of an intersection. If we fail to overcome this anisotropy, then sectioning the tissue will transfer the anisotropy to tissue sections and ultimately bias the probability that the planar surface or linear features in the tissue will make isotropic intersections with the geometric probes for surface area (test line) and length (test plane). For this reason, unbiased estimation of surface area and length requires special care to ensure that the intersections between the probe and features of biological interest are isotropic.

In the previous section we illustrated two of at least three options to overcome the inherent anisotropy of planar surfaces and linear features: (1) make the tissue isotropic (IUR sections) or (2) make the combination of probe and tissue isotropic (VUR sections for surface and VUR slices for length). The next section discusses the third and perhaps most versatile approach to date—using a 100% isotropic probe to estimate the length of linear features in tissue sections.

The randomization of the geometric probe prior to intersecting the tissue, rather than randomization of the tissue, allows tissue to be sectioned at any convenient, arbitrary orientation. However, either of these methods will ensure isotropic probe–object intersections, as required for theoretically unbiased estimation according to the Buffon principle. Option one is generally the preferred choice for biological tissue. The reason is that if the probe is isotropic, the tissue can be sectioned at any convenient orientation (e.g., coronal, longitudinal). Now that we have redefined the needle problem, its solution can be transformed into a biologically meaningful statement: *The probability of isotropic intersections between linear objects and 2-D planes is directly related to the length of the objects and inversely related to the distance between the planes.*

IUR and VUR sections focus on randomization of the object in the tissue. For IUR sections, all axes of the tissue are randomized; for VUR sections, one axis is fixed and the remainder are randomized. The disadvantage of these methods is that the experimenter cannot select a preferred or convenient section orientation. As shown in the following sections, the advent of computer-assisted stereology systems permits randomization of the geometric probe, rather than the object in the tissue. Both approaches achieve the ultimate goal—random probe–object intersections. In the following section we discuss two approaches involving isotropic probes that allow a convenient section orientation to be selected by the investigator.

Virtual Isotropic Planes

Isotropic planes are virtual planes generated at random angles by computer software. These virtual planes allow one to sample a fixed volume within a thick tissue section. Because the orientation of the isotropic planes is randomized in 3-D, random intersections occur between the surface of the planes and the linear structures of interest. A theoretical approach using isotropic virtual planes has been proposed for the estimation of L and L_V (see Larsen et al., 1998). Though clever and theoretically unbiased, the effectiveness of this approach is limited by two practical disadvantages. First, a significant amount of training is required for the observer to learn to keep track of objects and complex counting rules on sections viewed in multiple focal planes. Second, generation of isotropic planes requires computer software; to date, manual approaches have not been successfully developed. The method is important from the historical perspective as the first example of probe randomization for estimation of L and L_V.

Virtual Isotropic Spheres (Space Balls)

The idea for using a sphere to estimate linear features of stereological interest was clearly stated by Gokhale in 1990, with practical applications following about a decade later. Because the surface of a sphere is perfectly isotropic, all intersections of linear features with the surface have an equal probability. Based on the probability-based solution to the needle problem, the isotropic surface of a sphere provides an unbiased probe for making length estimates of linear features in tissue sections cut at any convenient orientation (coronal, horizontal, sagittal, etc.). This is because the requirement for isotropic probe–feature intersections is ensured by the isotropy of the probe rather than the feature in the case of IUR sections.

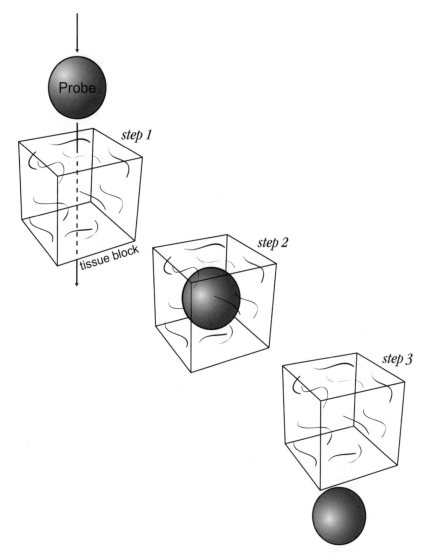

Figure 8-8. Schematic 3-D representation of the three-step process for scanning linear features in tissue using a virtual sphere probe. (From Mouton et al., 2002.)

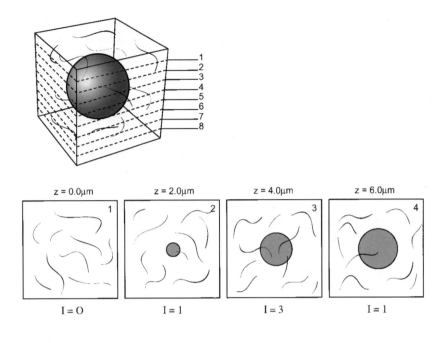

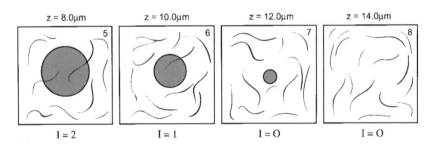

Figure 8-9. Schematic representation of isotropic sphere probe showing intersections (*I*) at eight planes through the z-axis of tissue containing thin linear fibers of interest

Figure 8-8 provides a schematic overview of scanning linear structures using an isotropic sphere probe. The first step shows the sphere probe about to pass in the z-axis of a tissue block containing linear features of biological interest (fibers, filaments, axons, capillaries, etc.). In the second step, as the operator focuses in the z-axis, the virtual sphere is revealed within the tissue section. In the third step the virtual sphere has "passed" through the tissue block. As shown in Figure 8-9, in practice the sphere does not actually pass through the tissue; rather, as the operator focuses on optical planes in the z-axis, the sphere is revealed in the form of circles (plane through a sphere = circle). The process can be done using grid overlays either affixed to a video monitor or generated over the image using computerized software (see Calhoun and Mouton, 2001). The operator indicates each time a linear feature intersects the sphere's surface and, according to the classic formula of Smith and Guttman (1953), the number of sphere–feature intersections is proportional to the length of the linear feature. For further details of this mathematical relationship, see Gokhale's proof in the appendix in Mouton et al. (2002).

One caveat applies to all methods using isotropic probes, either planes or spheres, to estimate length based on intersections with linear features: The higher the ratio of probe surface area to the diameter of the linear feature, the less stereological bias is present in the estimate. That is, not all geometric possibilities are included in the assumptions present in all stereological methods for estimating length based on intersections between planar probes and linear features of interest. Although the reasons for this are beyond the scope of this book, the basic issue is that such methods for estimating length are based on intersections between probes and imaginary spines in the center of the linear feature, rather than the linear feature itself. For linear features with certain geometries, this assumption is weak, although these geometries are indeed rare in biological objects. Nevertheless, the user should be aware of these limitations when applying modern stereological methods of length estimation to actual biological tissue.

In the case of isotropic sphere probes, one way to increase the ratio of probe surface to feature diameter, and thereby reduce this potential bias to negligible levels, is to divide the sphere into hemispheric probes or quartospheric probes. Because these probes require less space, they can be placed in tissue sections with thicknesses as high as 15 μm while reducing the bias from counting imaginary spines to less than 5% (Mouton et al., 2002). Although they are less efficient than a sphere probe, two times the number of hemispheric and four times the number of quartospheric probes achieve the same precision as one sphere probe. The surfaces of these probes are isotropic nevertheless and therefore lead to theoretically unbiased estimates of length based on the formula of Smith and Guttman (1953). In such cases we can use these probes with confidence that although stereological bias is not 100% eliminated, the results from the method will be unbiased for all practical purposes.

Figure 8-10 shows a series of optical planes in the z-axis. All points on the surface of a sphere have an equal probability of intersecting objects in the 3-D space. When the sphere passes through a section of tissue containing the

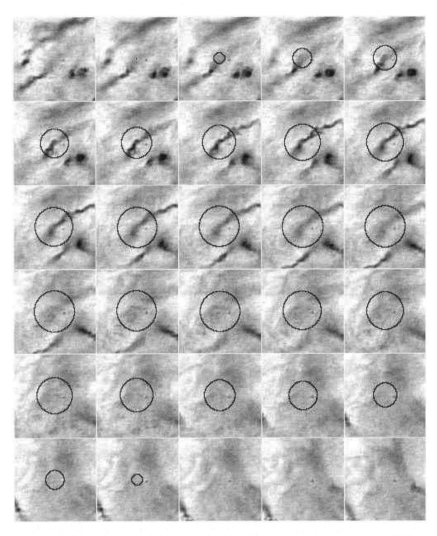

Figure 8-10. Estimation of total object length using virtual sphere probes focused at 0.2-μm steps in the z-axis of tissue sections containing thin linear fibers of biological interest

biological fibers of interest, the sum of sphere–fiber intersections with the surface of the sphere is proportional to the linear dimension of the fiber. Like the random toss and the factor $2/\pi$ in Buffon's principle, using a sphere to probe length and surface area provides an effective means to randomize the probability of probe–object intersections.

The same formulas described for estimation of length (L, L_V) using planes and lines apply to the use of isotropic spheres. The fractionator approach provides perhaps the most straightforward method for scaling local estimates of L_V to total L. The procedure is carried out on a known fraction of sections, in a known fraction of the reference area on each section, and in a known fraction of the section thickness. A fourth fraction, the probe sampling fraction (psf), is needed to estimate linear features using spheres that occupy a known fraction of a 3-D sampling box. This ratio is described as V_{sphere} / V_{box}.

$$\text{probe sampling fraction} = V_{sphere} / V_{box}$$

where $V_{sphere} = 4\pi r^3/3 = \pi d^3/6$ (r is the sphere's radius, d is the sphere's diameter) and $V_{box} = \text{length} \times \text{width} \times \text{height}$. Thus the complete formula for estimating total length using sphere probes and the fractionator method is

$$L = 2 \times \Sigma\, \mathrm{I} \times F_1 \times F_2 \times F_3 \times F_4$$

where $F_4 = 1/\text{probe sampling fraction} = 1/V_{sphere}/V_{box}$.

In this strategy, the number of intersections between the isotropic surfaces of spheres in a known fraction of the total reference volume is proportional to the total length for the total reference volume. As described previously, the fractionator approach avoids the unequal changes in the dimensions of the numerator or denominator that can occur using the two-stage method ($L = L_V \times V_{ref}$). The method using isotropic spheres applies equally well to the two-stage approach and will be theoretically unbiased provided that L_V and V_{ref} are estimated on final histological sections, that is, after all tissue processing is complete.

Virtual Cycloids for Estimation of Total Surface

Recently a method has been developed for making theoretically unbiased estimates of total surface area in a defined reference space, without the need to rotate tissue around at least one axis. This approach uses "virtual cycloids" to probe surfaces of interest within tissue, and requires computer software to generate cycloids with their major axis parallel to the VUR axis. The novelty of this approach is that any direction of sectioning can be used as the vertical axis. In contrast to VUR sections in which tissue must be rotated around a single axis, with the virtual cycloid approach, the user is free to select any sectioning orientation as the vertical axis, or to use sections that have been previously cut in any convenient orientation, without introducing stereological bias into sample estimates. With the development of virtual cycloids, theoretically unbiased estimates of total surface area can be achieved on tissue sections cut at any convenient orientation. This is similar to the development of isotropic probes for theoretically unbiased estimates of length in biological tissue.

Virtual Cycloid Ribbons for Total Length

Another virtual probe has been developed for estimation of total length on tissue sections too thin (<10 μ) for estimating length using isotropic sphere probes. The probe is a cycloid line projected along a second dimension, the so-called virtual cycloid ribbon. In contrast to length estimation using IUR sections and VUR slices in which tissue must be rotated around at least a single axis, the use of virtual cycloid ribbons allows for length estimation on archival sections and sections cut at any convenient orientation. As with the isotropic sphere probe and the virtual cycloid probe discussed earlier, any direction of sectioning can be used as the vertical axis. In practice, appropriately programmed computer software orients the virtual cycloid ribbon in the z-axis of the section, with the major axis of the cycloid ribbon parallel to the vertical axis. In terms of efficiency, the virtual cycloid ribbon approach is superior to isotropic virtual planes, but less efficient than isotropic sphere probes.

Summary

Modern stereological approaches take advantage of probability theory to estimate first-order parameters of biologically interesting objects. As illustrated by the Buffon principle, theoretically unbiased stereological estimates are based on an observable event—the intersection between a geometric probe and the object of interest. Isotropy (equal in all dimensions) is an important concept for theoretically unbiased estimation of orientation-sensitive parameters (S and L). The inherent anisotropy of biological objects must be overcome to ensure random probe–object intersections for theoretically unbiased estimates of surface area (S_V and S) and length (L_V and L).

There are three options to ensure isotropic probe–object intersections for estimation of surface area and length in tissue sections. First, the object in the tissue can be oriented at random using IUR sections; these sections are then probed with straight test lines. Second, intersections between an object and an isotropic probe (sphere) can be counted. In this case the section orientation can be selected for convenience by the investigator. Finally, the number of intersections between partially randomized objects (VUR sections) and partially randomized probes (cycloids) can be counted. To avoid the bias associated with ratio estimators, local estimates of L_V and S_V are scaled to the total reference volume using either the two-stage approach or the fractionator method.

As is the case for other stereological parameters, surface area and total length are dimensional quantities that can be affected by biological and artifactual sources of shrinkage and expansion. In the case of biological causes of changes in these parameters, the purpose of the experiment may be to document these changes. However, as indicated in earlier chapters, estimates may be affected by tissue-processing artifacts. Usually it is difficult to verify the extent to which differential shrinkage or expansion and other nonbiological artifacts affect the linear and surface area parameters of biological objects. Rather than attempting to estimate the effects of such artifacts, then somehow derive formulas to compensate for them, it is probably safest to try to diminish their effects.

We have discussed here the importance of scale in estimates of surface area and length. Thus, although we have described several theoretically unbiased approaches for estimating these parameters, it helps to keep in mind that stereological estimates apply strictly to a particular resolution and a specific set of tissue-processing conditions. Bias arising out of nonstereological sources can be referred to as nonstereological bias, as discussed in detail in Chapter 9.

9

Nonstereological Bias

---

*This chapter focuses on nonstereological sources of bias that can intro-
duce systematic variation into morphometric data. As discussed in
Chapter 2, true stereological bias cannot be measured, reduced, or eliminated; when present, it permanently prevents accurate estimates of
biological parameters. In contrast, nonstereological bias can be reduced
or eliminated. Examples include a variety of staining artifacts, including poor penetration, nonspecific antigen–antibody binding, and incorrect dilution. As factors causing nonstereological bias are identified,
removed, and/or minimized, sample estimates will converge on the
expected value of the parameter.*

The purpose of morphometry (quantitative morphology) as applied to
biological structures is to make sample estimates without introducing systematic error (bias), which causes these estimates to vary from the true or expected parameter. Bias refers to the difference between the mean or central
tendency of a sample estimate and the true central tendency of the parameter. Because theoretically unbiased stereology uses sampling and estimation
approaches that have been specifically designed to avoid all sources of systematic error, modern stereology is frequently referred to as design based, to
distinguish its approaches from those of assumption- or model-based stereology. The focus of Chapter 2 was on stereological sources of bias, including
assumptions, models, and correction factors. In this chapter we review the
major sources of nonstereological bias that can introduce systematic error
into sample estimates.

Sources of Nonstereological Bias

Bias from all sources can lead to systematic error. However, we can draw a
striking and significant distinction between the two types of bias. Stereological bias arising from assumptions and models leads to *inaccuracy;* once pre-

sent, nothing can be done to measure, diminish, or remove this inaccuracy. In contrast, sources of nonstereological bias can be identified, minimized, or eliminated entirely. Stereological bias arises from theoretical sources, while nonstereological bias arises from the practical application of stereological methods to biological tissue.

Once nonstereological sources of bias have been identified in a pilot study, steps can be taken to minimize or completely eliminate them. For instance, processing can be modified to allow greater penetration of stains into tissue; recognition of objects and reference spaces can be improved; and samples can be extrapolated to the correct population.

It is outside the scope of this book to provide an in-depth review of the numerous sources of nonstereological bias one encounters in the wide variety of morphological studies. Instead, we provide examples of the types of systematic errors that can arise from nonstereological biases. By being aware of these potential sources of nonstereological bias, researchers using theoretically unbiased methods should be able to identify and minimize error in sample estimates.

The following sources can introduce uncertainty into sample estimates of a stereological parameter:

Operator bias
Tissue-processing bias
Ascertainment bias
Recognition bias

Operator Bias

Operator bias is introduced when the operator (the person collecting data) favors one outcome over another.

Example: When tests are being performed to determine if treatment X causes cell loss in a defined reference space, a cursory examination reveals cell loss in some regions while other regions show no apparent cell loss, and still other regions show an apparent increase in cells. Oversampling of regions showing cell loss, which amounts to undersampling of regions showing no cell loss, will cause a systematic error (overestimation) in estimates of cell loss. This nonstereological bias favoring one outcome over another is called operator bias. Note that operator bias can favor either the desired or the undesired outcome.

One method for minimizing operator bias is to collect data blind, that is, without knowledge of the sources being evaluated. Blind data collection ensures that all parts of a reference space and all objects within the reference space have the same probability of being included in the data. In most cases blind data collection is easily achieved by using opaque tape to cover identifying information on a slide. Occasionally a tissue will contain identifying features that preclude strictly blind procedures. In these cases, data collectors should be especially aware of the potential for operator bias.

Tissue-Processing Bias

Tissue-processing bias results from artifacts introduced by the processing of tissue for stereological analysis. Objects of interest in the tissue are not fully visualized or cannot be unambiguously identified.

Example: When testing the hypothesis that treatment X reduces the average length of linear objects in a defined reference space, the reference space is cut into sections at an instrument setting of 50 μm and stained using immunocytochemistry on free-floating sections. The investigator determines that the post-processing section thickness is about 20 μm (about 60% shrinkage). A pilot study shows that the primary antibody penetrates only 5 μm from both sides of the tissue; thus objects in the middle 10 μm of each section are unstained. No further stain penetration can be achieved despite further manipulations of the procedure (e.g., heating, longer reaction times, a different primary antibody). Analysis of sections despite this caveat will lead to systematic error (under-estimation) of total length.

Most of the shrinkage or expansion in the thickness of tissue sections occurs through the movement of water. For example, water evaporates as tissue dries prior to application of a cover slip. In some cases changes in the thickness of sections can ensure better stain penetration. For instance, if a stain penetrates 15 μm from each surface of a 50-μm thick section, then cutting sections at an instrument setting of 25 μm should allow linear structures to be stained throughout the entire section thickness. However, all steps in the tissue-processing procedure can introduce nonstereological bias. Regardless of the cause, the effect is the same—the objects of interest cannot be unambiguously identified. In Figure 9-1, one can recognize a number of nonstereological biases. For these free-floating stained sections, the immunocytochemical stain did not penetrate through the full thickness of the section. In this case sampling would necessarily be limited to the regions at the top and bottom of the section, where full staining was complete. Before estimating stereological parameters, all sections should be examined closely to ensure the absence of tissue-processing bias.

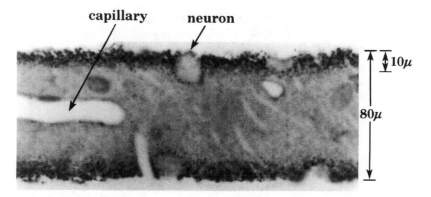

Figure 9-1. A tissue section cut at 80 μm and viewed perpendicular to the direction of staining shows incomplete penetration of immunoreactivity through the z-axis of the section. (From Calhoun et al., 1996.)

Ascertainment Bias

Ascertainment bias is systematic error arising when results are based on analysis of one population and extrapolated to another.

Example: Human tissue is frequently collected on the basis of unusual features. Stereological examination of tissue under the assumption that the sample reflects the full range of pathology (i.e., tissues with and without unusual features) can lead to systematic error (overestimation) for the unusual features. Analyzing a parameter in the biased sample with overrepresentation of the unusual feature would introduce ascertainment bias into the sample estimate.

The problem in this example is that the sampling is directed at one population and the interpretation of the results at another population. We might predict underrepresentation of samples showing all possible morphologies and an overrepresentation of samples with unusual features. If the goal is to extrapolate interpretations to the entire population, then efforts are needed to sample a representative portion of this population. If a population-based sample is not possible, then the conclusions should be limited to the actual population analyzed.

Recognition Bias

Recognition bias is systematic error introduced when the operator(s) cannot unambiguously identify whether the objects of interest fall within the defined reference space.

Example: Using the optical fractionator method, one moves through tissue and counts the number of objects (e.g., cells) according to the disector's unbiased counting rules. When a cell falls within a disector, the operator must be able to tell whether the cell is part of the reference space. Thus it is not necessary to draw distinct borders around a reference space; rather, it is necessary to know when a particular object that is being sampled belongs in the reference space. Either over- or underestimates will result when the operator cannot consistently identify whether cells are part of the population of interest.

Avoidance of recognition bias requires that the operator have a clear understanding of the morphological characteristics of the objects and reference space of interest. At the outset of the study, all persons involved should spend time reviewing the sampled and stained sections that are to be quantified. Examination of individual sections will reveal a set of morphological features that define the objects of interest and the reference space. These characteristics create consensus among the persons doing the study and will be the features reported to other investigators in publications.

A specific type of recognition bias arises from optical artifacts. As shown in Figure 9-2, light (arrow) passing through a tissue section (A) may cause the viewer to perceive a variety of artifacts (B).

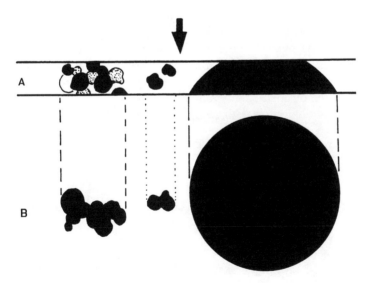

Figure 9-2. Sources of nonstereological bias arising from artifactual viewing of objects in tissue (A) and as projected images (B)

Inter- and Intra-Rater Differences

When multiple operators are collecting data, or when new operators are being trained, one way to ensure low recognition bias is to have different persons analyze and compare results for the same set of sections (*inter*-rater reliability). The similarity of these results will depend on whether different investigators are using the same criteria for the borders of the reference space and recognition of the objects of interest. For studies with sufficient sampling to ensure low sampling error, the estimates of different raters should fall within 10% of each other. Similarly, multiple estimates by the same operator on the same sections should be within 10% of each other (high *intra*-rater reliability). In cases of low inter-rater and/or low intra-rater reliability, more effort is needed to ensure that operators use consistent criteria for recognizing and counting the same population of objects.

In Situ versus *Ex Vivo* Parameters

An important distinction exists between estimates of number distributions and estimates of other parameter distributions of biological structures, including volume, surface area, and length. In the case of number, provided stereological and nonstereological sources of bias are avoided, sample estimates refer to the true distribution for the total number of objects in a defined reference space—the total number in the natural or unaltered state, for all the organisms in the population. For parameters other than number, sample estimates are designed to estimate the *ex vivo* parameter distributions, that is, the parameter as it exists on tissue sections, after all tissue processing is complete. In the case of total volume and mean object volume, the *ex vivo* volume distributions cannot be assumed to be equivalent to the true volume distributions. The reason for this differential is artifactual. Changes in the volumes of tissues and objects are caused by at least three factors:

> Postmortem events (autolysis, putrefaction, degradation)
> Temperature (freezing, heat)
> Reagents (fixative, stains)

Biological processes, such as postmortem events, can affect the transition between *in vivo* and *ex vivo* states. The classic example in the field of gerontology described in Chapter 2 has essentially eliminated the use of density estimators (parameter per unit of reference space) for comparisons of morphological change. An important consequence of this artifact for aging studies is that it invalidates the assumption of constant changes in the reference space for organisms of different ages. These assumptions affect estimates of numerical density (object number per unit of volume), but not of total volume.

The magnitude and direction of artifactual changes in volume are specific to the particular protocol in each study. Based on careful studies that have examined the effects of postmortem events and processing artifacts on tissue (fixatives, reagents), some volume distributions appear to be affected more severely by these artifacts than others. For instance, estimates of total volume in tissue are affected to a greater extent than local estimates such as mean cell volume. As indicated earlier, these artifacts are primarily caused by the movement of fluids, particularly water, between adjacent physiological compartments and between compartments and the environment. Water comprises a greater proportion of extracellular volume than intracellular volume. In biological objects bounded by a cell membrane, physicochemical forces may act as a barrier to the free movement of water, thus reducing the effects of postmortem changes and tissue-processing artifacts on mean object volume. As a result, a greater percentage of tissue volume is vulnerable to artifactual changes in the volume distribution during the transition from *in vivo* to *ex vivo* states.

Methods for direct estimation of volume include direct determination of volume and snap-freezing. The immersion method based on the Archimedes principle allows the fresh volume of tissue to be determined and compared with that after tissue processing:

1. Fill a container with water and place it on a scale.
2. Weigh the water + container.
3. Suspend the object by a thin thread in the water.
4. Subtract the weight of water + container.
5. The result equals the volume of the object in cubic centimeters.

Another method for assessing the effects of tissue fixation is snap-freezing. Snap-freezing fresh tissue using isopentane cooled by Dry Ice preserves the composition of tissue; however, this approach drastically reduces the quality of histological preparations compared with tissue fixation. The effects of postmortem events on the volume distribution of biological tissues are more difficult to assess. In theory, this problem can be approached using methods to image tissue *in vivo*, estimate volumes from stacks of images using the Cavalieri method, then compare these estimates with the volume of tissue in the fresh state prior to processing. However, this approach is complicated by imaging artifacts and the wide variety of agonal changes that occur in tissue during the transition from living to postmortem conditions.

Because of the effects of processing on parameter distributions, volume and other non-number parameter estimates are directed at the postprocessing parameters of the reference space as it appears on tissue sections. This caveat should not be misinterpreted as support for number as the only parameter that can be estimated without bias. Rather, the distribution of object numbers *in vivo* is the only *in vivo* parameter distribution that can be estimated without bias. For all other parameters, the *ex vivo* distributions are estimated without bias using the methods described in this chapter and the next. These methods use the same guiding principles from stochastic geometry and probability theory to make theoretically unbiased estimates of population parameters. The difference lies in the definition of the parameter distribution being estimated. For volume, surface area, and length, the central tendency and variation of the *ex vivo* distribution are the parameters of interest.

Standardization of the processing protocol for tissues in the same study avoids within-subject and between-subject bias from processing artifacts. Investigators should be mindful of this caveat when designing protocols to harvest and process tissue for later comparisons between sample estimates. If tissues within groups and especially across groups are subjected to similar protocols, the resultant variation from this source will be random, as required for statistical comparisons and the theoretical unbiasedness of *ex vivo* volume estimates. In this regard the dependence of volume estimation on protocol is analogous to the scale (resolution) dependence of total surface area and total length, as detailed in Chapter 8.

Summary

The use of design-based methods is the first step toward making theoretically unbiased estimates of biological structures. The critical second step is the application of theoretically unbiased methods to practical situations. Because of the wide range of applications possible when estimating first-order stereological parameters, we have not attempted to provide a comprehensive discussion of all possible nonstereological sources of bias. Rather, the investigator should carefully identify and eliminate all potential sources of error that can introduce nonstereological bias. In those cases where nonstereological sources of bias cannot be fully eliminated, the investigator must decide whether the effect of the bias is significant. If the decision is made to allow the bias to remain, it is conventional to provide a thorough description of the bias to allow readers the opportunity to agree or disagree with the results.

10

All Variation Considered

Thus far this text has emphasized sampling and geometric probes for the estimation of first-order stereological parameters. In this chapter we examine the important issue of variation, a second-order parameter. In contrast to morphometry that focuses on classically shaped objects, which are by definition devoid of variation (e.g., "assuming a cell is a sphere . . ."), the stochastic geometry and probability theory underlining modern stereology takes variation into account. When successfully applied to biological tissue, the variation observed in parameter estimates using unbiased stereology can be partitioned into two sources: sampling error and biological variation. While the latter quantity arises from natural forces outside of the investigator's control, the former is determined by the sampling frequency chosen by the investigator. By selecting a sampling frequency that leads to a sampling error well below the biological variation, the investigator ensures a sampling design that is optimized for maximum efficiency.

Estimates versus Measurements

Imagine that your goal is to compare the functional significance of this book with that of other books. The first task is to identify the basic functional unit of a book, which is a sentence. The purpose of a book is to transmit ideas in the form of letters, words, and phrases, all combined into meaningful sentences. The sentences in a book are combined into paragraphs, paragraphs into pages, and pages into chapters. Paragraphs contain a higher order of functional significance since they are composed of related sentences; levels of organization below the sentence transmit only partial ideas. In comparison, there is little functional significance to the number of pages and chapters; they simply divide the book into convenient units. Every sentence ends in a period, or some variation thereof (?, !); thus, we can view the number of periods or sentence endings as a reliable indicator of the number of basic functional units in a book. If we count the number of periods in every paragraph, on every page, and in every chapter, we will know exactly the number of basic functional units in the book. We can think of this determination as a measurement. To compare different books, we could carefully repeat the measurement process. Finally, to compare different categories of books, for example in a particular library, we could count the number of periods in each type of book on the basis of genre (romance versus detective), publisher (university versus trade), publication date (old versus new), etc.

Or we could make systematic-random estimates. Instead of counting every period on every page, we could count periods on every tenth page, with a random start between pages 1 and 10. To estimate the total number of sentences in each book, we would then multiply the actual number of periods counted by 10. Of course, the estimated number of sentences would vary, depending on the location of the random start. However, the estimate would be theoretically unbiased by any false assumptions (e.g., all books have the same number of pages) and therefore would be accurate. Instead of counting every period on every tenth page, we could count periods on every fifth line on every tenth page, again with a random start between the first and fifth lines. To estimate the total number of sentences, we would multiply the number of actual periods counted by 5 for every fifth line, then 10 for every tenth page. For a book of 300 pages with approximately 40 lines on each page, we could make a reliable estimate of the total number of functional units (sentences) by simply counting periods on about 8 lines (every fifth) on each of 30 pages (every tenth), for a total of 240 lines per book. Since it takes about 30 seconds to count each page, the time per book for making a theoretically unbiased estimate of total sentence number will be about 30 sec/page $\times$ 30 pages = 900 sec = 15 min. In this way, systematic-random sampling can considerably shorten the time and effort expended to obtain information without sacrificing accuracy.

For a theoretically unbiased estimate of the number of functional units (sentences) within an entire category of, say, 200 books, we would not count periods in every book, but rather would sample every twentieth book in the category, again with the first book selected randomly between books 1 and 20. In this example we would estimate sentences for the entire 200 books from a systematic-random sample of 10 books (200/20 = 10). By counting the number of periods on every eighth line, on every tenth page, and on every twentieth book in the category, we are able to make a reliable and theoretically unbiased estimate of the total number of sentences for an entire category of books in less than 3 hr (15 min/book $\times$ 10 books = 150 min = 2.5 hr). To estimate the total number of functional units for the entire category, we would simply multiply the actual number of periods counted by 1000 [5 (lines) $\times$ 10 (pages) $\times$ 20 (books) = 1000] to obtain a theoretically unbiased estimate of the total number of functional units.

Because the purpose of the exercise is to make comparisons between groups of books, rather than the individual books themselves, such estimates provide an accurate and theoretically unbiased method for quantifying objects. More important, such an estimate can be done in a small fraction of the time and at a small fraction of the cost of an actual measurement. Because the time and cost saved can be used to estimate more objects in each group, systematic-random estimates are more efficient than actual measurements.

The goal of sampling in stereology is to make estimates in which the mean value reflects the expected value of a parameter. When applied to studies of biological structure, systematic-random sampling permits accurate (theoretically unbiased) estimation of morphological parameters. For each study, the actual item of interest will vary, depending on the biological question and the structure that carries the greatest amount of functional significance for the project. None of the individual values in an estimate is expected to be accurate; rather, the power of the estimate stems from the accuracy of its mean value, the mean total number of cells, the mean cell volume, or the mean total length of fibers. When assumptions about the size, shape, or orientation of the objects being estimated are minimized and if possible avoided, the objects can be arbitrary shapes, and the estimates are free from bias related to assumptions, models, and correction factors.

Measurements in biological studies differ from other estimates in several critical ways. The goal of a measurement is to make a single accurate determination of the value of a parameter. Examples include measurement of pH with a pH meter, measurement of body weight with a bathroom scale, or determining the volume of an object based on the amount of water it displaces (Archimedes' principle). Although mean values for measurements can be determined from a sample of individuals in a defined population, the focus is on the accuracy of each measurement. When the object of interest has a classical geometric shape, classical geometry permits its parameters to be calculated from a single measurement. For such approaches, the accuracy of the measurement depends on the accuracy of the measurement device and the ability of the person making the measurement to avoid errors. A measurement device's accuracy is usually reflected in terms of the smallest units that the device can reliably measure, such as ± 1.0 μm. However, the accuracy of a measurement should not be mistaken for the efficiency of a particular measurement device. The efficiency of a measurement device depends on only two factors: the expected range of the measurements and the time required for the measurement. For instance, a device accurate to ± 1.0 kg would be efficient for measurement of an object expected to be about 100 kg, too accurate for a 1000-kg object, and not accurate enough for a 100-mg object. Obviously the best way to pick an efficient measurement device is to first obtain a rough idea of the expected value.

Estimates, on the other hand, are designed to quantify the mean value of a parameter for a population of objects. When theoretically unbiased sampling and estimation methods are employed, the mean value of the sample estimate falls within the true distribution of the parameter at the population level. Here we can see how an estimate differs from a measurement. As discussed above, the accuracy of measurements is a function of how close the measured value is to the expected value. For sample estimates using unbiased methods, none of the individual estimates of a sample parameter are designed to hit the expected value; rather, accuracy is defined by whether the mean sample estimate is expected to converge on the expected value after further sampling. Thus, instead of the weight of a particular person, a theoretically unbiased estimate would provide an accurate estimate of the mean weight of the population to which that person belongs. In this case it will be less important to measure the exact weight of each individual to a high level of accuracy; instead, the most efficient strategy is to roughly estimate the weight of many individuals sampled at random from the population.

As discussed in this chapter, estimates make better sense when possible differences between populations are hypothesized. For instance, suppose you hypothesize that Democrats weigh more on average than Republicans. The efficient way to test this hypothesis is to make blind estimates of the weight of numerous Democrats and Republicans, sampled at random from the population, rather than taking excruciatingly accurate measurements of a few Democrats and a few Republicans. Thus, estimates are better for testing population differences; they capture the true variation in each population faster than measurements can.

Relative and Absolute Differences in Parameters

One frequently hears from biologists that they are not interested in the absolute differences in a parameter. Rather, they only want to know the relative differences (i.e., the difference between a control group and a treatment group), and therefore less time and effort should be required to obtain this information. With a little sober reflection, however, we can see that this is a distinction without a difference.

When we think of the relative group differences for a parameter, for example, a 45% difference between the treatment and control groups for the number of cells in a defined reference space, we are talking about the *true* relative difference. Moreover, we do not mean to imply that this relative difference applies only to the small number of individuals in the sample, but rather that the difference is true for the population from which the individuals were sampled. For individuals sampled at random from a population, there are two ways to find a mean value—either from measurements or from estimates. Of these, estimates provide a far more efficient approach to finding the mean value of a population parameter.

Now we have arrived at the crux of the problem. The goal of the exercise is to estimate the mean value of a parameter from a small sample of subjects selected at random from a population. The population can be members of the same species; they can differ in age, gender, or some other independent variable. There can be one population or two or more populations in the case of testing hypotheses and making inferences about treatment effects, In the latter case, if the sample estimates show a statistically significant difference in the parameter between one or more groups, then we can make inferences about the possible meaning of this difference, relative to the different treatments administered to each group. As this discussion shows, good stereology is simply a part of good experimental design. If the parameter estimates are reliable, then the conclusions drawn about the biological meaning of these estimates will be reliable as well.

Estimates and Hypothesis Testing

Frequently the goal of stereological studies is to test hypotheses related to differences between two or more groups. Stereology provides the best method available to obtain accurate estimates of morphological parameters because it can provide *sample estimates* of a *population parameter*. When we estimate parameters in several individuals from each group, we will find the mean value of the sample estimates for each group. The goal of hypothesis testing is to determine if these mean values are statistically different, that is, whether there is a strong likelihood that the means of the sample estimates reflect true differences for the population parameters.

The reader should note that not all stereological studies are designed to test hypotheses. Some studies may be designed to estimate the range of a morphological parameter in a group, for example, the total number of cells in a defined reference space from a certain species. In these cases stereology is used because the investigator wishes to make a reliable estimate of the parameter.

Technical studies are another use of stereological data that does not involve statistical testing. Technical studies may be designed to test the feasibility of combining stereological analysis with a new staining method or with a novel microscopic technique. Although the results may not be used for testing a hypothesis, it is important to follow proper stereological procedures since the results may be used for qualitative comparisons.

The majority of stereological studies are designed to test hypotheses about some aspect of biology involving morphological structure, for example, morphological changes in disease compared with healthy controls, the effect of treatment A versus controls. Therefore the first step of a stereological study is to have a clear idea of the hypothesis being tested. In hypothesis testing we begin with the null hypothesis, the idea that we hope to *disprove*. That is, if we think that a particular treatment A causes a reduction in total cell number (N) compared with a placebo, the null hypothesis is

null hypothesis: mean N (treatment A) = mean N (placebo)

With statistical testing we hope to reject this null hypothesis, and thus obtain support for the alternative hypothesis, that the treatment has a significant effect on cell number:

alternative hypothesis: mean N (treatment A) $\neq$ mean N (placebo)

Although the null hypothesis focuses on the central tendency or mean values of parameters in two or more groups, hypothesis testing in biological sciences uses the variation around these values as the basis for rejecting a null hypothesis. For this reason the variation in a sample estimate, rather than the mean value, frequently receives greater attention during a stereological study. In statistical terms, incorrectly rejecting a null hypothesis when it is true is called a type I or alpha statistical error. This error is an error of the worst kind because it adds misleading information to the scientific literature and wastes resources that could be focused on new discoveries. The use of theoretically unbiased stereology allows us to maximize the probability that our mean and variation estimates are accurate and thus avoid type I errors.

Optimization for Maximum Efficiency

At this point we can begin to optimize the sampling design for maximum efficiency. This process requires preliminary data from a pilot study, first with a single individual containing the reference space of biological interest, then with one or two individuals from each of the groups in the study. Optimization begins with the realization, shown by numerous studies during the past decade, that essentially any reference space of biological interest is ideally sampled by about 10 (range 8 to 12) systematic-random sections. Although the exact number of sections required to capture most of the variability in a particular reference space will be determined as the pilot study progresses, this range of sections through any reference space provides a starting point.

Analyzing fewer than this number may not ensure adequate sampling of the reference space, while as the number of systematic-random sections through any reference space exceeds 12, one passes the point of diminishing returns in the investment of time, energy, and resources.

The next step in optimizing a sampling design for maximum efficiency is to calculate the sampling interval that will generate the approximate number of desired sections. As indicated earlier, if the reference space is present on 100 sections, taking every tenth section with a random start from sections 1 to 10 will suffice. The ideal number of cells for a reliable count of total number is between 100 and 200 objects counted through the entire reference space. This ideal level of sampling can be achieved by selecting the appropriate spacing between disectors. In a pilot study one can estimate the approximate disector spacing (i.e., distance between disectors) needed to ensure that the equally spaced disectors fall within the reference space. Similarly, to space the sampling evenly throughout the reference space, the ideal number of disectors used to count the objects is the same as the ideal number of objects counted. Obviously this simplifies to counting one object per disector.

$$\text{number of cells per disector} = (100-200 \text{ cells}) \div (100-200 \text{ disectors})$$
$$= 1 \text{ cell per disector}$$

These study parameters are determined empirically for each stereological investigation by using a pilot study on one set of sections. If the pilot study shows that too many cells or too many disectors are counted for a given disector size and distance between disectors, one has a number of options to increase efficiency. If the pilot study on a single individual shows that the number of objects counted per disector is averaging more than two to three objects per disector, then the area of the frame and/or the height of the disector should be adjusted downward, and thus reduce the number of cells falling within each disector. Second, if one finds that for a given disector spacing, 300 to 400 disectors fall on sections containing the reference space, then increasing the distance (spacing) between disectors will reduce the number of disectors hitting the reference space. After these changes are made, a repeat of the pilot study on a single individual will indicate if they have increased the sampling efficiency. In practice, the coefficient of error, a measure of sampling error, provides a useful quantity for optimizing sampling efficiency.

The Pilot Study

A pilot study provides a tremendous amount of information to the investigator at an early point in a stereological study. Whenever possible, it is best to carry out pilot studies blind, and continue this practice throughout the study. That is, the person collecting the data should not be aware of which individual being analyzed belongs to which group. In addition to avoiding the possibility of introducing unconscious bias into the results at an early stage, blind collection strengthens the conclusions based on findings in the study.

If we use theoretically unbiased methods to make estimates for a defined reference space, then the mean value of the sample estimate will be accurate and *might* provide a reliable estimate of the mean parameter for the population. Whether the estimate is a reliable reflection of the population parameter depends on how much effort is used to capture the variation observed in the parameter; that is, how *precise* the mean estimate is. From statistics we know that precision refers to degree of variation. Precision is inversely related to variation. A mean estimate with high variation is less precise than a mean estimate with low variation. Thus, to increase the precision of an estimate, we must reduce its variation around the mean.

The pilot study provides a rough idea of the mean and variation of the sample estimate for each group. In a few rare cases, the differences between the means will be striking after only two or three individuals are analyzed. If the variation around each mean is low, there may be a statistically significant difference between group means. In this case a statistical test may allow the null hypothesis to be rejected, with the probability of a type I error being less than 5% ($P < 0.05$). Then an alternative hypothesis can be supported with a 95% or better chance of accuracy. However, in most cases the pilot study is simply the starting point for an analysis of statistical power—how many individuals should be analyzed to show a statistical difference (Chapter 12).

The Analysis of Variation

Typically the variation in a sample estimate will be high when fewer than five individuals are analyzed. In this case the low statistical power, that is, the small number of individuals in the study, limits the ability to discern statistically significant differences, if they exist. The next step is to optimize the sampling protocol for maximum efficiency. This ensures that most of the variation in the sample estimate is from biological variation rather than sampling variation. Because we are making estimates based on sampling only a part of the total tissue of interest, we can control the sampling intensity. We can chose to sample more or less of the tissue. Later we will describe how to *partition* the variation in a sample estimate into its component sources. Once we know the sources of the variation in a sample estimate, we can increase the precision of the estimate by sampling where the greatest variation is present. This process begins with the calculation of the total amount of variation in a sample estimate, also known as the total observed variation. Figure 10-1 shows the expected high level of total observed variation for counts of

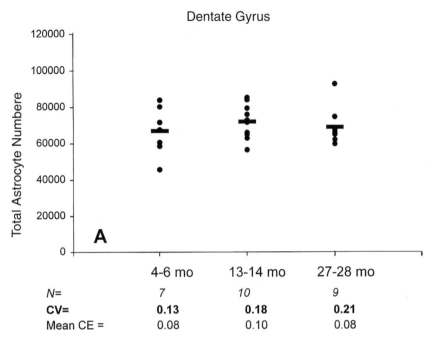

Figure 10-1. Results for total cell (astrocyte) number in a defined brain region (dentate gyrus) for mice of different ages. Each point signifies the value for a different individual. Note the high biological variability (true interindividual differences) compared with the relatively low sampling error (mean CE). (From Long et al., 1998.)

cells in a defined reference space (the mouse dentate gyrus). For theoretically unbiased stereological methods, the total observed variation can be partitioned into two sources: biological variation and sampling error.

Total Observed Variation

The first step in finding the total observed variation (total variation) is to determine the mean and standard deviation of the sample estimates. The mean value is simply the sum (Σ) of the individual estimates (I), divided by the number (n) of individuals analyzed: $\Sigma I / n$.

To find the variation of the individual estimates around this central tendency, we calculate the standardized or standard deviation, SD. The SD for the sample estimate is the square root ($\sqrt{}$) of the sum of the squared differences of each value (x) from the mean value (mean), divided by the number of individuals:

$$SD = \sqrt{[\Sigma(x - \text{mean})^2 / (n-1)]}$$

The SD for small samples, that is, fewer than thirty individuals, is calculated using $n-1$ individuals in the denominator. This formula is preferred in stereology because we typically have sample estimates based on fewer than thirty individuals per group in sample estimates.

Calculation of the mean and SD for a sample estimate leads to an important question: How much precision is needed? Although each of the individual estimates in a sample may be accurate, analyzing too few individuals may not capture enough variation in the parameter to provide a reliable mean estimate. With low statistical power, we risk failing to reject a null hypothesis that is in fact untrue, also known as a type II or beta error. Although type II errors are generally less serious than type I errors, failing to document a true statistical difference can be misleading in future studies. Traditionally, the minimum number of individuals needed to document a significant difference is five in each group. However, showing that a difference does not exist generally requires analysis of a larger number of individuals than showing that a difference exists. In either case, the first step in analyzing stereology results is finding the total variance in the estimates. Second, before analyzing more individuals following the initial pilot study, you should determine the source of variation (i.e., partition the variation) in your sample estimates.

The Coefficient of Variation

From the mean and SD we can calculate the total observed variation as the coefficient of variation:

CV = total observed variation = SD/mean

Because the number of individuals analyzed is present in the denominator for the SD and mean, the larger the sample size, the lower the expected CV. However, analysis of more individuals requires the expenditure of tissue resources, labor costs, supplies, and time. Therefore, the best strategy is to estimate the source of most of the variation in the sample estimate; that is, what proportion of the total variation stems from analyzing too few individuals (low statistical power) or low sampling within each individual. If low sampling within individuals contributes most of the variation in the sample estimate, then simply analyzing more sections, and/or more locations within each section, is the most efficient strategy to reduce the total variation.

An Example

Suppose we have an animal model in mice in which cells die within a defined reference space. Now we wish to test the hypothesis that a particular treatment A results in significant protection against cell loss in the model. We decide to test this hypothesis using theoretically unbiased stereology (the optical fractionator method) to estimate the total number of cells in mice receiving treatment A and those receiving a placebo. Because we hypothesize that the treatment prevents cell loss, we expect a lower number of cells in the placebo-treated group and a higher number of cells in the group receiving treatment A. Therefore, the first step is to design a pilot study to estimate total cell number in the reference space of interest. For pilot studies involving two or more groups, the sampling should be based on the group expected to show the lowest value for the parameter of interest, in this case the placebo-treated group. This sampling design, when applied to all the groups in the study, will result in sufficient sampling to estimate and partition total variance in the pilot study.

We can proceed with a pilot study of the placebo-treated mice to analyze the variation for the total number of cells. The first dataset in Table 10-1 is for 3 placebo-treated mice; the second shows data for these 3 mice, plus 2 additional mice from the same group. The total reference space was serially sectioned and 10 sections for each mouse were systematically sampled. A total of 100 disectors spaced in systematic-uniform-random locations throughout the reference space were sampled across all 10 sections.

We can see that increasing the number of individuals analyzed from 3 to 5 increases the precision of the estimate almost 40% (i.e., reduces the CV from 0.37 to 0.24). Of course this increased precision comes at a price in terms of materials, time, and other costs (labor, supplies, etc.), as shown in Table 10-2.

Cost factors can be included in the study design by considering the efficiency of each estimate in terms of precision (CV), as a function of the total time (hr) and total cost ($) for each dataset. As Table 10-2 shows, because of the larger number of individuals analyzed in dataset 2, its efficiency is more than four times lower than that of dataset 1. However, the added cost for dataset 2 is justified because of the considerable gain in precision from adding 2 mice to the study. Now that we have an estimate of the precision and cost factors, we can proceed with the pilot study.

Table 10–1 Relation between Sample Size and Statistical Power

	Dataset 1 (3 individuals)	Dataset 2 (5 individuals)
Number of sections per individual	10	10
Number of disectors per individual	100	100
Analysis time (hr) per individual	4	4
Cost per individual ($)	50	50
Mean number of cells ($\times 10^3$)	37.0	32.2
SD ($\times 10^3$)	13.6	7.6
Coefficient of variation	0.37	0.24

Table 10–2 Relation between Sampling Power, Precision, and Costs

	Dataset 1 (3 individuals)	Dataset 2 (5 individuals)
Total number of sections	30	50
Total number of disectors analyzed	300	500
Coefficient of variation (CV)	0.37	0.24
Total analysis time (hr)	12	20
Total cost ($)	150	250
(Efficiency [CV/time/cost]) ($\times$ 10^5)	21	5

We estimated the results for 5 mice from the placebo group during our pilot study. Now we add sample estimates for 5 mice from the treatment A group. The results are shown in Table 10-3. Inspection of the data suggests a possible increase in total number of cells in the treatment. A group compared with the placebo group. Next we test whether this apparent difference is statistically significant. Specifically, we will test whether we can reject the null hypothesis: total N (treatment A) = total N (placebo), and in doing so provide support for the alternative hypothesis: total N (treatment A) $\neq$ total N (placebo). We can do this using the Student t-test to test this hypothesis when the individuals are selected at random from the population (independent sampling, see Chapter 3); the parameter of interest has a unimodal, symmetric distribution and the two groups show similar variation around their respective means. The formula for the t-test is

$$t = \text{absolute difference between means} \div \sqrt{(\Sigma \text{var}^2 / n)}$$

Table 10–3 Comparison of Cell Numbers
in Sample Groups

Individual	Placebo (no. cells $\times$ 10^3)	Treatment A (no. cells $\times$ 10^3)
1	29	44
2	34	34
3	26	30
4	36	46
5	36	36
Mean	32.2	38.0
SD	7.6	6.8
CV	0.24	0.18

where Σvar^2 is the sum of the squared variances for two means (variance = SD^2).

Inserting the values from Table 10-3,

$$t = |38.0 - 32.2| \div \sqrt{[7.6^2 + 6.7^2)/5]}$$
$$= 5.8 \div 4.5$$
$$= 1.3$$

At this point we consult a table from a statistics text showing values from the t-distribution. From the t-distribution for the two-tailed t-test, we see that the t-value calculated from the dataset must be greater than 2.3 in order to reject the null hypothesis with a 95% level of confidence ($P < 0.05$) and 8 degrees of freedom $[(n = 5) + (n = 5) - 2]$. The t-value from our dataset, however, is only 1.3, not greater than 2.3, and therefore is not large enough to allow us to reject the null hypothesis. Had it been larger than 2.3, the alternative hypothesis that treatment A might prevent cell loss would have been supported. At this point no further analysis would be required and the study would be complete. Without a calculated value of t greater than 2.3, however, there are no statistically valid grounds to support the alternative hypothesis.

We can now reexamine the data in the sample estimates. The actual difference between the means of the two groups is about 5800 cells, or an 18% increase over the placebo group. If in our judgment the finding is that treatment A does not provide significant protection against cell loss in a small number of treated animals, then we may decide to report the data as a negative finding. Alternatively, if on the basis of the pilot study we think a significant difference may arise after more sampling, then a further analysis of the results from the pilot study is indicated.

Sensitivity of the Test Statistic

In the above example we could not reject the null hypothesis because the calculated t-value was less than the target t-value of 2.3 for 8 degrees of freedom at a $P < 0.05$ level of confidence. If we examine the t-table, we see two strategies to lower the target t-value and thus increase the chance of rejecting the null hypothesis. One strategy is to find a target t-value at a lower level of confidence. However, the conventional thinking in biological research is to use a 95% confidence level ($P < 0.05$) as the cutoff for statistical significance. A level of $P < 0.05$ means that we are willing to accept a 5% risk that the null hypothesis will be falsely rejected when it is in fact true. Notice that using this conventional approach, we expect that in 1 out of 20 studies the null hypothesis will be incorrectly rejected. Nevertheless, bioscientists are willing to accept this level of uncertainty. Rejecting the null hypothesis with only a 90% confidence level ($P < 0.10$) would allow one-tenth of all studies to incorrectly reject the null hypothesis. Thus, lowering the level of confidence to 90% would double the level of uncertainty in the scientific literature.

A second approach for decreasing the target *t*-value would be to increase the degrees of freedom (df). As shown by inspection of a table of values from the *t* distribution, the higher the degrees of freedom, the lower the *t*-values for the 95% level of confidence. Before considering this option further, we should review what is meant by a "degree of freedom." Every time we take an average or mean value for a set of data points, we "lose" one df. For the mean value to accurately represent the dataset, all of the numbers are free to vary *except the last one.* For instance, 5.9 is the mean of 3.1, 5.6, 2.2, 13.7, and 4.9. This statement is true for this dataset if and only if the last number is 4.9. If the last number is any other number, the mean of the five numbers cannot be 5.9. Thus, taking a mean value requires placing a constraint on one of the numbers in the dataset. For this reason, the total df for a comparison of two means is always equal to the total number of values in the dataset, minus 1 for each mean that was taken. For the example comparing the mean values for two groups with $n = 5$ per group, the total df was 8 (5 + 5 − 2).

As more values are included in the mean value, the effect of losing the df diminishes and the target *t*-values in the *t*-table decrease. We say that a dataset with more values has greater *statistical power* than a smaller dataset. As shown in a table of *t*-values, as the df increases, the *t*-value decreases, thereby increasing the probability of rejecting the null hypothesis at the 95% level of confidence. Thus we see that one clear way to increase the chance of rejecting the null hypothesis is to add values to the dataset; that is, to analyze more individuals. However, as we saw earlier in this chapter, analyzing a greater number of individuals is associated with higher cost in terms of time, effort, and material. To avoid the possibility of wasting these resources, before we decide to increase statistical power by analyzing more individuals, we should also consider less expensive, faster, and less laborious methods that could reduce the total variation and lead to a statistical difference between the groups.

Recall that we base our decision to reject the null hypothesis on the calculated value of *t*, which is equal to the ratio between two quantities: the difference between the means and the square root of the pooled variance between groups. Here again is the formula for the *t*-statistic with values from our example:

$$t = \underbrace{|38.0 - 32.2|}_{\substack{\text{difference} \\ \text{between means}}} \div \underbrace{\sqrt{[(7.6^2 + 6.7^2)/5]}}_{\substack{\text{pooled} \\ \text{variance}}}$$

The pooled variance in the sample estimate is determined by the intensity at which we sampled the reference space. That is, by sampling a greater proportion of the reference space within each individual, we could reduce the pooled variance in the sample estimates and thus increase the calculated *t*-value. This source of error (variation) is related to how the tissue is sampled and is sometimes referred to as sampling error. If we sample a given reference space with a low intensity, then the sampling error will be high. However, the level of sampling error is only one of two factors that contribute to the pooled variance. As indicated earlier, besides sampling error, total variation includes biological variation, which is the true difference in a parameter between different individuals in the population. No amount of sampling within individuals can reduce biological variation. Thus, we see the two possibilities for a high pooled variance in sample estimates:

Low statistical power (low *n*)
Insufficient sampling within each individual

We now have identified two possible strategies for increasing the probability of rejecting the null hypothesis. The first strategy involves increasing the statistical power by analyzing more mice in each group. Here we hope to show a true difference between means by sampling more individuals from each group, thereby increasing the numerator in the *t*-statistic. The second advantage of increasing *n* is lowering the target *t*-value by increasing total df. Although it is expensive in terms of resources, this could favor rejecting the null hypothesis by increasing the numerator in the *t*-statistic, thus lowering the pooled variance, and by lowering the target *t*-value. The second strategy is to reduce the sampling error. This would be relatively inexpensive because it would involve further sampling of the tissue that is already available. However, for this option to succeed, the sampling error would have to make a relatively large contribution to the total variation. As reviewed in the next section, we use the data from our pilot study to estimate the relative contributions of sampling error and biological variation to the total pooled variance in the sample estimate.

Sampling Error and Biological Variation

Recall that we obtain the mean and SD for each group from the results of our pilot study. The CV ($CV = SD/$mean) is a mathematical expression for the total observed variation in a sample estimate. The two factors that contribute to the magnitude of the CV are the biological variation (BV) and the sampling error (CE). Biological variation arises from the interindividual differences for a parameter in a reference space. An example of biological variation would be the difference in brain volume for mice of a particular age and strain. Despite our best efforts to control all known factors, we can expect to find variations in the parameter from one individual to the next. Biological variation is present regardless of how intensely we sample the reference space. In contrast, sampling error stems directly from the intensity that we use to sample the reference space.

In testing a hypothesis, the BV is the most important source of variation. It provides the true variation between individuals that is the basis for analyzing possible group differences. Because the BV is defined at the level of the population rather than the sample, we cannot determine it; that is, not without knowing the value of interest for every individual in the population. Fortunately, we can estimate the BV from the two other quantities, the CV and CE, both of which can be derived from the sample estimates in a pilot study. Having an estimate of BV permits us to optimize the sampling scheme in the study for maximum efficiency and thus make the best use of our time, material, and labor resources.

Partitioning the Total Observed Variation

We partition the total observed variation (pooled variance) in sample estimates to determine the most efficient approach to further sampling. First, however, before adding and subtracting sources of variation such as CV, BV, and CE, we must square these quantities to express them as variances: CV^2 (total observed variance), BV^2 (biological variance) and CE^2 (sampling variance). Because the sampling variance denotes the mean of the CE^2 values for each group's sample estimates, we refer to the observed sampling variance for each group as the mean CE^2. These variances are related as shown in the following equation:

$$CV^2 = BV^2 + \text{mean } CE^2$$

Solving for BV,

$$BV^2 = CV^2 - \text{mean } CE^2$$
$$BV = (CV^2 - \text{mean } CE^2)^{1/2}$$

In this way BV can be estimated from the total observed variance [$CV^2 = (SD/\text{mean})^2$] and the observed sampling variance in the sample estimates for a small number of individuals in a pilot study. In the following section we discuss how to estimate the observed sampling variance (mean CE^2).

Observed Sampling Variance

Consider a study to quantify the total number of cells in a defined reference space. However, instead of an estimate of the total cell number obtained by sampling the reference space with a theoretically unbiased stereological probe (disector), suppose we count all the cells; that is, we make a determination. In this case there is no sampling error because there is no sampling. When we partition the variation for a determination of this type, 100% of the observed variance (CV^2) is attributable to biological variance. Mathematically the expression would be

$$CV^2 = BV^2$$

Now imagine that we realize that it is unnecessary to make a determination of the parameter; an estimate will do just fine. As we move from a determination to an estimate, we progressively increase the amount of sampling in the reference space, and hence increase the sampling error. Remember that the accuracy of determination is ensured by the use of a theoretically unbiased method. Provided that we also sample in an unbiased manner, we can also ensure the accuracy of our estimate. As shown in Figure 10-2, the more sampling we do, the more variation we expect for each estimate.

The figure shows the effect of sampling error on the range of expected values for an estimate of total number (*est N*) using a theoretically unbiased method, for example, the disector principle. The horizontal line shows the expected value of the parameter, about 66 objects. This number is contained within a total reference space consisting of 96 sections. The y-axis plots values for the estimate of the total number of objects that would be expected for the decreasing levels of sampling indicated on the x-axis. Starting at the far left on the x-axis, the reference space could be sampled by 16 samples of 6 sections each, 12 samples of 8 sections each, and so on, progressing to 1 determination on 96 sections as shown on the far right. When all 96 sections are used, there is no sampling, no sampling error, and only a single value (solid horizontal line), which is the expected value of the parameter in this individual. The principle of sampling error illustrated in Figure 10-2 applies for all estimates of biological structures using modern stereological methods, such as systematic-random sampling and unbiased geometric probes, for the four first-order stereological parameters. As sampling of the reference space increases, the estimate progressively converges on the true or expected value; this is the stereologist's definition of an unbiased estimate.

Based on this discussion, one could conclude that a determination is preferable to an estimate if one is interested in a reliable value for a particular parameter. At this point, however, we are advised to recall Weibel's adage to do more less well, which applies to all estimates of stereological parameters. As discussed in the following sections, there are two excellent reasons why an estimate is preferable to a determination. The first, as mentioned earlier, is that there is a wide range of expected values for a parameter across dif-

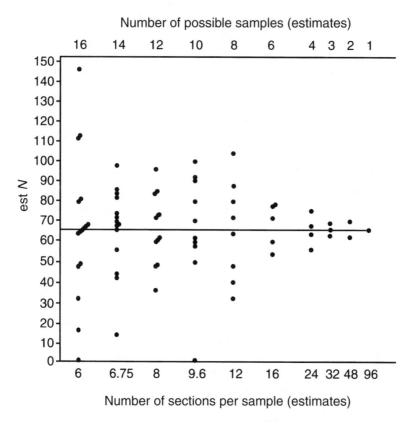

Figure 10-2. Expected values for the estimate of total number of objects (est *N*) following analysis of different numbers of sections sampled in a systematic-uniform-random manner. The horizontal line at 66 indicates the true value expected for counting all cells on all 96 sections. (From West, 1999.)

ferent individuals in a population (high interindividual variation or biological variation). It is simply not good use of resources to make highly precise determinations in each individual when high biological variation causes wide variation among values for individuals in the same population. Second, the efficiency of systematic-random sampling allows the total variation of an estimate to be captured by a relatively small sample of the population of interest.

Interindividual Variation (Biological Variation)

Because of natural selection, we do not expect all individuals in a defined population to show exactly the same value for a particular parameter. Instead we expect a range of values, with the magnitude of the range regulated by evolution, that is, selection pressure from forces outside of the population. As an example, consider the expected range of values for your body temperature after a night of sleep. As long as your body is maintaining a normal level of homeostasis, the value of this parameter should be 37°C (98.6°F) or within one or two degrees of this number.

Thus, we say that there is low interindividual variation in body temperature. This narrow range of values for body temperature defines the optimal activity of literally millions of physiochemical reactions required for normal bodily function. The optimal body temperature has a central tendency at 37°C (98.6°F), as opposed to 5°C (41°F) or 27.2°C (81°F). Furthermore, as individuals in this population, we are under constant selection pressure to maintain the narrow range of this parameter. If one morning we awake with a body temperature outside of this range, for example, a fever of 40°C (104°F), this value will lie considerably outside the narrow normal range. Should we be unable to effectively identify and overcome the cause of the abnormal body temperature, then within a relatively short time we, as well as our abnormal body temperature, will be removed from the population, thus restoring the narrow normal range.

Biological forces tend to converge individual estimates on the central tendency for the parameter in the population. The level of biological variation around this central tendency is a function of the selection pressure exerted on the parameter. Figures could be obtained for all individuals in a population of interest. From these figures we would observe, similar to Figure 10-2, two independent sources of variation for each individual. One would be the biological variation, that is, the differences between the horizontal lines indicating the range of expected values for the parameter in different individuals. Second, there would be the spread of the values on the y-axis that is entirely the result of sampling error. The combination of this biological variation and the sampling error would be reflected in the total observed variation in our sample estimate of the parameter in this population.

From the viewpoint of efficiency, we might ask: At what optimal level of sampling do we make reliable estimates without wasting resources by oversampling too many individuals or sampling too much within each individual? The answer requires some preliminary knowledge of the level of biological variation in the parameter. That is, once we suspect that the interindividual variation of values within each group is X percent, we can then set a much lower target value for the optimal level of variation within each individual, say one-half or X percent.

Returning to the example given earlier, we can partition the variances for the sample estimates from the placebo (control) and treatment A groups as follows.

The placebo group:
CV $= 0.24$; CV$^2 = 0.24^2 = 0.0576$
CE $= 0.15$; CE$^2 = 0.15^2 = 0.0225$
BV$^2 = 0.0576 - 0.0225 = 0.0351$
BV $= \sqrt{0.0351} = 0.187 = 0.19$

Treatment A group:
CV $= 0.18$; CV$^2 = 0.18^2 = 0.0324$
CE $= 0.12$; CE$^2 = 0.12^2 = 0.0144$
BV$^2 = 0.0324 - 0.0144 = 0.018$
BV $= \sqrt{0.018} = 0.134 = 0.13$

This process of partitioning variance shows that BV^2 accounts for only about 60% of the observed variance in the placebo group [(0.0351/0.0576) $\times$ 100 = 0.60] and about 56% [(0.018/0.0324) $\times$ 100 = 0.56] of the observed variance in the treatment A group. The remaining variance in the sample estimates, about 40 to 45% for both groups, is sampling error. Therefore, increasing the sampling intensity within each mouse will substantially reduce the pooled variance, which will increase the calculated t-value, and thereby increase the probability of rejecting the null hypothesis.

Reducing Sampling Error

As shown above, when sampling error is high relative to the biological variance, the optimal strategy for reducing the pooled variance is to reduce the sampling error. By setting the sampling error at about one-half or less of the biological variation, one can be sure that only a negligible amount of the total variation in the sample estimate stems from sampling, with most of the variation attributable to true interindividual differences (i.e., biological variation). Sampling error can be reduced by increasing the number of sections analyzed, increasing the sampling within each section, or both. The choice among these options depends on whether within-individual factors, the between-section variation or the within-section variation, contribute most strongly to the sampling error.

A look at Table 10-1 reveals that the sample estimates in both treatment A and the placebo groups are based on cell counts in about 100 systematic-randomly placed disectors throughout each reference space. In their classical 1987 paper, Gundersen and Jensen showed that most of the within-section variance can be captured by sampling at 100 to 200 locations. This study shows that sampling beyond this optimal level achieves little in terms of increasing the precision of the sample estimate, but rather wastes resources that would provide a greater benefit if used to sample more individuals. This concept is what the axiom, "do more less well" states in absolute simplicity.

One approach to reducing the sample variation in the example is to increase the sampling intensity; that is, sample the reference space with a greater number of sections, disectors, or both. Systematic-random sampling involves sampling biological objects along a single axis (dependent sampling) as opposed to sampling objects at random (independent sampling).

To understand the sampling error for dependent sampling, it is useful to consider its two component sources of variation in greater detail. One source is the variability related to the number of sections analyzed, the between-section variation, which arises from variation in the parameter, for example, from section 1 to section 2, from section 2 to section 3, and so on. Recent theoretical work from Professor Hans Juergen Gundersen's group at the Stereology Research Laboratory in Denmark indicates that most of the variation in any stereological parameter for any reference space is contained within a minimum of 6 to 8 systematic-random sections (Gundersen et al., 1999). As shown in Table 10-1, the sample estimates for treatment A group and the placebo group are based on an analysis of 10 sections through the reference space of each mouse. Thus, sampling more than 10 sections would not be expected to result in a significant reduction in sampling error.

For systematic-random (dependent) sampling, one of the two within-sample factors contributing to CE is the biological "noise," also know as the between-section variation or systematic-random-sampling (SRS) variation, which arises from variation in the parameter within each section. The second contribution to CE is variation in the parameter arising from sampling within each section, also known as within-section variation or the nugget effect. Taking the square of variation denotes *variance,* and allows for mathematical operations on these quantities:

$$CE^2_{\text{dep}} = \text{between-section variance (SRS)} + \text{within-section variance (nugget)}$$

The most current theoretical work shows that within-section variance rather than between-section variance makes the greater contribution to the sampling error. For the dependent, systematic-random sampling that is intrinsic to modern stereological approaches, the level of sampling error is inversely proportional to n, the number of objects counted:

$$\text{sampling error variance (dependent sampling)} = CE^2_{\text{dep}} \approx 1/n$$

This relationship shows that to reduce the sampling error by half, for dependent sampling one should count twice the number of cells, or in this example, increase the number of disectors counted from 100 to 200. We would predict that this increased level of within-section sampling would result in a dramatic reduction in the CE—from 0.15 to about 0.07 in the placebo group and from 0.12 to about 0.06 in the treatment A group. In both cases the increased level of within-section sampling would decrease the mean CE to less than half of the BV.

For independent sampling, the CE is equal to $CV/\sqrt{n}$. In this case the CE is inversely related to the square root of the number of objects counted:

$$\text{sampling error variance (independent sampling)} = CE^2_{indep} \approx 1/\sqrt{n}$$

Thus, for independent, nonsystematic sampling, a one-half reduction in sampling error would require counting four times as many disectors through the reference space. We can conclude, therefore, that dependent sampling along a single axis captures a greater proportion of the total variation per unit of time and therefore lends greater efficiency to the sample estimate.

The expected effect of increased within-section sampling on rejection of the null hypothesis is shown in Table 10-4. Increasing the number of disectors counted through the reference space from 100 to 200 will reduce the pooled variance, the denominator of the t-statistic. Doubling the number of disectors only reduces the pooled variance $[\sqrt{(\Sigma var^2/n)}]$ from 4.53 to 3.45, which is to say, not much. Despite high within-section sampling using 200 disectors and high between-section sampling using 10 sections through the reference space, the calculated t-value is still 1.7 ($t = 5.8/3.45$). Recall that a t-value of at least 2.3 is required to reject the null hypothesis for 8 df at a confidence level of $P < 0.05$. From this analysis we can conclude that the major source of variation in the sample estimate is biological variation, not sampling error. Because BV arises from interindividual differences, this analysis shows that only sampling more individuals can produce appreciable reductions in the total variation of the parameter.

This example illustrates one possible scenario that may arise when partitioning variance from a sample estimate. The high total variance in the sample estimate prevented rejection of the null hypothesis in the original pilot study with five mice each in the treatment A and placebo groups. By examining the contribution that both sampling error variance and biological variance made to the total variance, we determined that reducing the sampling error might reduce the total variance and lead to a value of t high enough to

Table 10–4 An Example of the Effect of Increasing Within-Section Sampling on the Calculated t-Value

Sampling frame	Difference betrween means	$\sqrt{\Sigma var^2/n}$	t-value
100 disectors	5.8	$\sqrt{[7.6^2 1 6.8^2)/5]}$	1.3
200 disectors	5.8	$\sqrt{[(5.2^2 1 5.7^2)/5)]}$	1.7

reject the null hypothesis. By partitioning the variance, we found that sampling error variance (CE^2) contributed between 40 and 45% of the total variance, with the remainder arising from biological variance. By further dividing this sampling error variance into its components, we found that the within-section variance (the nugget effect or stereological noise) arising from a low number of disector locations accounted for most of the sampling error. By increasing the number of disector locations from 100 to 200, we reduced the CE^2 to less than half the biological variance. Nevertheless, despite this reduction in the total variance in the denominator of the t-statistic, the calculated t-value failed to exceed the value required to reject the null hypothesis. Based on this analysis, we can conclude that further reductions in the total variance will require analysis of more individuals. If we analyze two additional mice instead of doubling the number of disector locations, we could predict the following results for 100 disectors on 10 sections:

	Difference between means	$\sqrt{(\Sigma var^2/n)}$	t-value
100 disectors	5.8	$\sqrt{[(7.6^2 + 6.7^2)/7]}$	3.83 ($P < -0.005$)

These results show that a slight increase in statistical power permits rejection of the null hypothesis with a 99.5% level of confidence. Although the cost of analyzing two additional mice is greater than analyzing 100 more disectors in the tissue sections on hand, the above analysis shows that in the particular scenario presented here, analyzing more mice is the only rational strategy available. Because sampling error variance made only a small contribution to the total variance compared with the large contribution of the biological variation, increasing the number of sampling locations (disectors) within each section had a minor effect on the total variation compared with increasing the number of individuals analyzed.

This example shows that once the sampling error in a sample estimate is sufficiently low relative to the biological variation, increasing the statistical power may be the only way to reduce the total variation in a sample estimate. This strategy cannot reveal a statistical difference that is not present, but rather helps show a difference that is true. A word of caution, however: Adding df (i.e., increasing statistical power) reduces the target t-value required to reject a hypothesis; in theory, with enough statistical power, any null hypothesis can be rejected. For this reason we distinguish a biologically significant effect (i.e., rejecting a null hypothesis with a low number of individuals) from a statistical effect related to unusually high statistical power.

Variation of Assumption-Based Estimates

In this chapter we have shown how analysis of variation from sample estimates using methods that reduce systematic error to a negligible level permits an experimental design to be optimized for maximum efficiency. It should be emphasized that partitioning total variance into biological variance and sampling error variance pertains exclusively to sample estimates obtained by theoretically unbiased methods. The sampling schemes and estimation probes used in modern stereological methods are specifically designed to limit error in sample estimates to random error arising from biological sources and sampling error. The variation in sample estimates from assumption- and model-based (biased) stereological approaches cannot be similarly analyzed. The reason is that sample estimates from biased methods contain the same biological variance and sampling error that is found using the approaches described in this chapter, as well as unknown quantities of variance from models, assumptions, and correction factors. Because these latter sources of variance cannot be quantified, we have no way to deduce the magnitude of their contribution to the total variance.

A mathematical analysis of variance for estimates from biased methods includes the term e to denote bias (systematic error) arising from models, assumptions, and correction factors:

$$CV^2 = BV^2 + CE^2 + e$$

As for estimates from theoretically unbiased methods, we rearrange and solve for biological variance:

$$BV^2 = CV^2 - CE^2 - e$$

Because e adds an unknown, unmeasurable, and nonremovable source of systematic error to the total variance in the sample estimate, it is not possible to determine biological variance with the level of certainty required for statistical analysis.

One Last Word about Efficiency

In the beginning of this book we mentioned that accuracy was the most important consideration in a stereological study, followed by precision and efficiency. In some cases, theoretically unbiased stereological approaches are less efficient than assumption-based morphometric methods. However, the efficiency of an estimate is not related to its accuracy. The ideal sample esti-

mate is one that uses theoretically unbiased methods, excludes nonstereological sources of bias, and captures most of the biological variation in the
parameter in a minimal amount of time.

Efficiency is expressed as precision and time ($E = P$/time). The efficiency of the estimate increases as the time to make an estimate decreases
and as the precision of the estimate increases. Sample estimates are designed to capture most of the variation in a parameter, which may or may
not be efficient, depending on how much variation is present. Therefore the
level of precision associated with a particular sample estimate is dependent
on the biological variability of the parameter. The precision of estimates of
first-order stereological parameters is expressed in terms of a second-order
stereological parameter (variation) that can be estimated directly from
sample estimates: the total observed variation ($CV = SD$/mean) and the
sampling error ($CE = CV/\sqrt{n}$). These sources of variation are inversely related to precision: as CV and CE decrease, precision increases ($P = 1/CV$; $P
= 1/CE$). Squaring these variations to express them as variances, we can ex-
press efficiency in terms of precision and time for total observed efficiency
and sampling efficiency:

total observed efficiency $= E_{CV} = P$/time $= 1/CV^2 \times$ time
sampling efficiency $= E_{CE} = P$/time $= 1/CE^2 \times$ time

Here we see why theoretically unbiased stereological approaches are frequently *less* efficient than assumption-based morphometric methods. Because the goal of modern stereology is to make theoretically unbiased estimates of the expected first- and second-order stereological parameters (i.e.,
mean and standard error), sample estimates may require more rigorous sampling and estimation procedures than assumption-based morphometric approaches. For a theoretically unbiased estimate, the full reference space must
be sampled and parameters reported in terms of absolute, first-order parameters. The less rigorous approaches of conventional morphometry afford
greater opportunities for taking shortcuts, avoiding sources of variation, and
sampling less than complete reference spaces, all of which reduce the time required for the estimate.

Second, because assumption-based methods are not designed to capture
all of the variation in a parameter, they frequently report less variation and
therefore more precision than theoretically unbiased estimates. In contrast,
the rigor inherent in theoretically unbiased methods increases both the time
required to make the estimate *and* the variation in the sample estimate, which
both lead to reduced efficiency compared with assumption-based (biased)
morphometric approaches.

Summary

Because theoretically unbiased stereology is designed to estimate the full variability in a parameter, partitioning variance permits sampling to be optimized for maximal efficiency. The first step in optimization is a pilot study in which the total variation in a parameter is estimated in a few subjects. From these estimates one can estimate the two possible sources of variability inherent in theoretically unbiased methods, biological variation and sampling error. If the sampling error accounts for more than half of the biological variation, it is efficient to reduce the observed variation by sampling more within each individual. However, if biological variability adds more than twice the contribution of sampling error to the total variation, the most efficient strategy is to sample more individuals from the population. The ability to partition variances among component sources is a critical advantage of theoretically unbiased stereology compared with assumption- and model-based stereological methods.

11

Typical Stereology Designs

This chapter presents some typical experimental designs that are used for analyzing biological structures. Perhaps the two most common stereological parameters estimated by biologists are total number of objects and total volume of tissue, followed by number-weighted mean object volume, total object length, and total object volume. While each study differs in its specifics, the six designs in this chapter provide the basics for most stereological studies.

Total Object Number

Purpose: To estimate the total number of objects in a defined region using the optical disector and Cavalieri methods, that is, the two-stage ($N_V \times V_{ref}$) method (Sterio, 1984; Gundersen, 1986; for biological applications, see West and Gundersen, 1990; Mouton et al., 1994, 1997).

Stereology: The total number of objects in a defined reference space can be estimated as the product of N_V (numerical density) obtained using the theoretically unbiased optical disector method (Gundersen, 1986) and a theoretically unbiased estimate of the reference volume, V_{ref}, obtained by using the Cavalieri estimator and point counting (see Gundersen and Jensen, 1987). Numerical density and reference volume were both estimated using theoretically unbiased (assumption- and model-free) approaches. Because the denominator in N_V and the value of V_{ref} both refer to the same volume, any changes in the reference space during tissue processing (i.e., shrinkage, expansion) will cancel in their product; therefore $N_V, \times V_{ref}$ will be a theoretically unbiased estimate of total number, N.

Tissue preparation: Serially section the reference space containing the objects of interest at an instrument setting of 40 to 50 μm. Sample a total of 8 to 12 sections in a systematic-uniform-random manner. Stain the sampled sections to visualize the objects of interest. Measure section thickness, t, at a total of 30 random locations on all sections at high magnification [100× (NA 1.4) oil-immersion objective].

Procedure: Identify a unique counting item within the object of interest. For example, for cells with a single nucleus, select the nucleus or nucleolus as the counting item for the estimation of total cell number. Count the number of objects, Q^-, in the reference space using 15.0-µm-high disectors (disector frame area = 975 µm²) with a guard distance of 3 µm from the cut surface of each section. Count the item when it appears in best focus while slowing focusing in the z-axis through the height of the disector. Adjust the dimensions of the disector (i.e., the area of the disector frame and the disector height) so that on average one to two objects are counted per disector. To increase the precision of the estimate of numerical density (number of objects per unit of disector volume, N_V), repeat this procedure at 100 to 200 disector locations at systematic-random locations throughout the reference space.

Formula: Calculate the total object number, N, as the product of numerical density, N_V, and the reference volume, V_{ref}:

$$V_{ref} = \Sigma P \times a(p) \times T$$

where:

V_{ref} = total volume of the reference space
ΣP = sum of points hitting the reference space
$a(p)$ = area per point (µm²)
T = distance between sections (µm)

The equation for numerical density is

$$N_V = \Sigma Q^- / (\text{number of disectors} \times \text{volume of one disector})$$

where ΣQ^- is the sum of objects counted, the number of disectors is the number of disectors counted through the reference space, and the volume of one disector (µm³) = a(frame) × disector height.

To obtain the total reference volume, estimate V_{ref} from the Cavalieri formula. From the estimates of total numerical density (N_V) and the reference volume, calculate the total number of objects, N, in the reference space from:

$$N = N_V \times V_{ref}$$

Precision: To determine if a sufficient number of disectors were sampled, calculate the sampling error:

$$CE = CV / \sqrt{n}$$

where CV is the total observed variation = SD/mean, and n is the number of individuals analyzed.

A sampling error of 10% or less (CE $\leq$ 0.10) is the desired target for biological studies. The ideal number of systematic-uniform-random sections through any reference space is 8 to 10 sections. Sampling more than this number in any given individual does not appreciably reduce the sampling error. If the CE is greater than 10%, and at least 8 sections have been sampled through the reference space, then further reductions in the sampling error (i.e., increases in within-sample precision) are most efficiently achieved by counting more disectors through the reference space rather than by analyzing more sections (see Gundersen et al., 1999). To count more disectors, decrease the spacing between disectors. Once enough sampling has been done to achieve a sampling error of 10% or less, sampling more individuals from the population of interest is the next most efficient strategy to further reduce the observed variation (CV).

This design was used by two groups, Mouton et al. (1994) and Ohm et al. (1997), to make theoretically unbiased estimates of the total pigmented cell number in a defined reference space (nucleus locus coeruleus) of the human brain (see Figure 4–6). These results are in strong agreement: They show little, if any, nonsignificant age-related cell loss in the nucleus locus coeruleus of the human brain.

Total Reference Volume

Purpose: To estimate the total volume of a defined reference space using point counting and the Cavalieri principle (Gundersen and Jensen, 1987; for a recent biological application to brain, see Mouton et al., 1998).

Stereology: The total volume of the reference space is proportional to the sum of the areas on systematic-uniform-random slices through a defined reference space (Cavalieri, 1635).

Tissue preparation: Remove the entire volume of tissue containing the reference space and section it in a convenient plane. Set the thickness of the slices to divide the reference space into 8 to 12 slices. Make the first coronal cut at a random point in the first interval and continue sectioning at uniform intervals throughout the entire reference space.

Procedure: Randomly orient a point grid over each slice or an image of the face of the section. Select an area per point for the grid that allows between 150 and 200 points to hit the entire reference space. Place the grid at random over the reference space on each slice, and count the total number of points hitting the reference space on all slices. Begin with the first section through the reference space and continue to the end in serial order. Calculate V_{ref} using the Cavalieri formula:

$$V_{ref} = \Sigma A \times \text{average } t \times k$$

where:

average t = average section thickness (μm)

k = sampling interval (e.g., every sixtieth section)

ΣA = the sum of the reference area on each section (μm²) = $\Sigma P \times$ area per point (μm²)

$\Sigma P \times$ area per point = the product of the total number of points hitting the reference area on each section (ΣP) and the distance between points in the x-direction and the distance between points in the y-axis (area per point)

Precision: To determine if a sufficient number of points and sections were sampled, calculate the sampling error ($CE = CV/\sqrt{n}$, where $CV = SD/$mean). A sampling error of 10% or less (CE $\leq$ 0.10) is a desirable target for biological studies. If the sampling error is greater than 10%, and at least 8 sections have been sampled through the reference space, further reduce the sampling error by counting more points, that is, by decreasing the area per point. Once enough sampling has been done to achieve a sampling error of 10% or less, to further reduce the CV, sample more individuals from the population.

This method was used to quantify the total cortical brain volumes for patients with Alzheimer's disease (AD) and for aged controls (Figure 11-1). When the points hitting the cerebral cortex on each section are plotted by their distance from the frontal pole of the brain, the effect of cortical atrophy in AD can be appreciated. The values for total cortical volume (V_{CTX}) plotted by age at death are shown in Figure 11-2. Finally, total cortical volumes in AD cases $\oplus$ and aged controls ($\Diamond$) can be correlated with severity of dementia (Mini-Mental State Examination), as shown in Figure 11-3.

Mean Object Volume

Purpose: To estimate the number-weighted, mean object volume (MOV) of objects in a defined reference space using the rotator method and vertical sections (Baddeley et al., 1986; Jensen Vedel and Gundersen, 1993; for biological application, see Mouton et al., 1994).

Stereology: The MOV is proportional to the average line length originating at a fixed point and projecting in random directions to the border of the cell profile. The vertical section design randomizes the estimate of average β across all possible line lengths through the object. Therefore the rotator method in conjunction with vertical sections provides a theoretically unbiased estimate of MOV. When the method is applied to a population of ob-

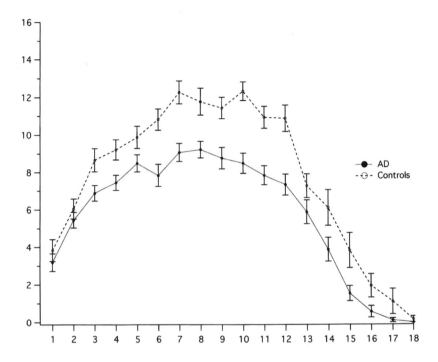

Figure 11-1. Total volume estimate using the Cavalieri-point counting method for human brain cortex in aged controls and Alzheimer's disease (AD) cases. At autopsy, AD cases show about 25% loss of total cortical volume compared with nondemented aged controls. (From Mouton et al., 1998.)

jects sampled using the optical disector, the resulting estimate of MOV is number weighted. Thus the estimate of MOV is theoretically unbiased for the true number distribution for the population of objects.

Tissue preparation: Dissect the entire reference space from the surrounding tissue. Identify the dominant long axis of the objects of interest as the vertical axis. Rotate the block containing the objects of interest around the vertical axis. Section the block at an instrument setting of 40 μm in the plane perpendicular to the axis of rotation. Stain the resulting vertical-uniform-random sections to reveal the objects of interest.

Procedure: To obtain a number-weighted estimate of MOV, the estimates of object size must be done on objects counted using a theoretically unbiased counting method, that is, the theoretically unbiased optical disector. For each counted object, select a fixed point in the object, such as the nucleolus. Orient three lines

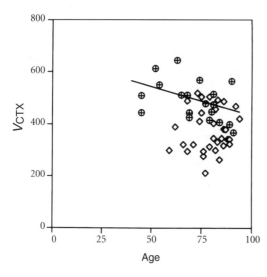

Figure 11-2. Total cortex volume in Alzheimer's disease (◊) and aged controls (⊕) as a function of age at death

over the object (four if there is high variation in object size) perpendicular to the vertical axis. Measure the length of each line from the fixed point to the border of the object. Repeat for 100 to 200 objects in the reference space.

Formula: Calculate the MOV of the population of cells according to the formula:

$$\text{mean object volume} = \text{average } l^3 \times 4\pi/3$$

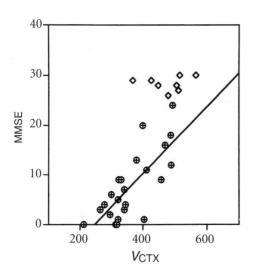

Figure 11-3. Strong direct correlation between cognitive impairment (Mini-Mental State Examination score, MMSE) and reduced total cortical volume at the time of death for Alzheimer's disease (⊕), but not for aged control (◊) cases

where average l^3 is the average of the cubed line lengths from the vertical axis through the object to the border of the object.

Precision: Calculate CE and CV for MOV. Increasing the objects sampled will reduce sampling and increase precision of the MOV estimate. Once the sampling error is 10% or less, further reductions in total variation are most efficiently achieved by sampling more individuals from the population.

Total Length

Purpose: To estimate the total length, L, of linear objects in a defined reference space using vertical-uniform-random slices (Gokhale, 1990; for biological applications, see Stocks et al., 1996). For an efficient variation applied to arbitrary (convenient) sections, see Mouton et al. (2002).

Stereology: The number of random intersections between linear fibers of unknown length and the 2-D probe in a known fraction of a reference space is proportional to the total length of the fibers in the reference space. A 2-D probe is created by moving a line probe through a known distance in the z-axis. The use of VUR slices and cycloids (sine-weighted lines) ensures random intersections between the linear features of interest and the 2-D probe. Note that the same result can be achieved using the surface of isotropic spheres (space balls) as the 2-D probe for linear features. Because the surfaces of spheres contain all possible orientations in 3-D space, VUR slices are not required; therefore the tissue containing the linear features of interest can be sectioned at any convenient orientation.

Tissue preparation: Dissect the reference space of interest from the surrounding tissue. Identify the dominant long axis of the linear objects of interest as the vertical axis, and rotate the block around the vertical axis. Orient the block to be sectioned in the plane perpendicular to the axis of rotation. Section each reference space exhaustively at an instrument setting of 40 to 50 μm. Using systematic-uniform-random sampling, sample about 8 to 12 sections from the total number of sections containing the reference space. Stain the sample of VUR slices to reveal the objects of interest.

Procedure: The number of sampled sections expressed as a fraction of the total number of sections represents the section sampling fraction. On each section use a sampling frame of known area [area(frame)] to sample a known fraction of the total area on each section (area of the xy-step = distance of the x-step × distance of the y-step). In the z-axis of each section, orient a 3-D sampling probe (disector) a known distance from the top and bottom of each section. The height of the disector expressed as a fraction of the total section thickness represents the thickness sampling fraction. Within the volume of each disector, orient a grid of cycloids of known total length with their major axis perpendicular to the vertical axis of rotation. While moving the focal plane slowly in the z-axis, count the total number of intersections (I) between the linear feature of interest and the cycloid line probes. Adjust the cycloid length within each sampling frame so that the sum of intersections (ΣI) for the total reference space is between 100 and 150.

Formula: Calculate the total length of the linear feature, total L, from the formula:

$$\text{total } L = \Sigma I \times a/l \times F_1 \times F_2 \times F_3$$

where:

ΣI = sum of the intersections between the linear feature and cycloids
F_1 = $1/ssf$
F_2 = $1/asf$
F_3 = $1/tsf$
a/l = area of the sampling frame per unit of length of the cycloid (μm)

Precision: Estimate CE and CV as described above, with a CE of about 10% or less as the desired level of sampling error. Sampling 8 to 10 sections through the reference space will capture most of the between-section sampling error, while sampling a greater number of sections will not produce an appreciable reduction in CE (Gundersen et al., 1999). To further reduce the CE, increase the number of sampling locations within each section by reducing the spacing between disectors.

Total Volume by Area Fraction

Purpose: To estimate the total volume fraction of objects from single random sections through a reference space. For example, consider a blood vessel containing intraluminal pathology of unknown volume.

Tissue preparation: Slice the reference space (blood vessel) containing the objects of interest into thin sections (<10 μm).

Stereology: Point counting can be used to estimate the area of an object as a fraction of a reference area. According to the Delesse principle (1847), for a random section through a reference space containing objects of interest, the point fraction [the ratio of the sum of points hitting object profiles to the sum of points hitting the reference area ($\Sigma P_{obj} / \Sigma P_{ref}$)] is proportional to the area fraction [the ratio of the profile area to the sampled reference area (A_{obj}/ A_{ref})], which in turn is proportional to the volume fraction [the volume of the object divided by the volume of the reference space (V_{obj}/V_{ref})], as shown in the formula:

$$\Sigma P_{obj} / \Sigma P_{ref} = A_{obj} / A_{ref} = V_{obj} / V_{ref}$$

Procedure: Outline the reference area on each section. In a systematic-uniform-random manner, move a frame containing a grid of points with known area per point [area(point)] to 100 to 150 x,y locations through the reference space. Adjust the area per point so that about 150 to 200 points hit profiles through the entire reference space. At each x, y location, count the number of points (P_{obj}) that hit profiles of the object and the number of points that hit the reference area (P_{ref}).

Formula: Calculate the total object volume as the product of the point fraction and an estimate of the total reference volume using the Cavalieri volume.

$$\text{total object volume} = (\Sigma P_{obj} / \Sigma P_{ref}) V_{ref}$$

Precision: Calculate the CE and CV as described above, with a CE of about 10% or less as the desired level of sampling error. Sampling 8 to 10 sections throughout the reference space will capture most of the between-section sampling error, while sampling a greater number of sections will not produce an appreciable reduction in CE (Gundersen et al., 1999). To further reduce the CE, increase the number of points hitting object profiles by decreasing the area per point; second, increase the number of disector locations by decreasing the spacing between disectors.

12

Frequent Questions about Stereology

The advent of theoretically unbiased stereological approaches has forced a number of philosophical changes in the way a biologist approaches the task of quantifying biological structures. Whereas in the past, morphometric analysis of biological structures depended heavily on the subjective expertise of the highly trained anatomist, modern stereological approaches focus on the accuracy of theoretically unbiased geometric probes and the unbiased efficiency of systematic-random sampling. This shift in emphasis has introduced major changes in the state-of-the-art methods of morphometry, from the techniques used to process tissue to the interpretation of results. As a result, biologists working with new stereology often have specific questions about the methodology. This chapter presents some of the most frequently asked questions and attempts to answer them.

Is a Pilot Study Necessary?

Unless a previous study has been done on the same parameter in the same reference space, a pilot study using two or three individuals serves several important functions. First, it enables one to optimize the sampling at all levels based on an empirical estimate of the true biological variation in the parameter of interest. This information indicates how to focus additional sampling, and hence time, effort, and resources, where it is likely to produce the most significant reduction in observed variation. Second, a sample estimate from a few individuals allows a rough determination of the number of individuals that should be analyzed to achieve statistical significance. The third asset of a pilot study is the opportunity to optimize tissue preparation procedures for maximal visualization of the biological objects of interest. For instance, stereological analysis through a tissue section requires complete staining of the features of interest. In the case of immunocytochemical procedures, full immunoreactivity through every section may require several trial-and-error experiments to find the optimal dilution, time in primary antibody, counterstaining, etc. Therefore, at an early stage in the process, a pilot study provides the information needed to optimize sampling parameters; it allows a prediction of the statistical power needed to complete the study and it provides the opportunity to modify the tissue-processing protocols for maximum recognition of structures.

Do Theoretically Unbiased Estimates Relate to *in Vivo* Structure?

Generally speaking, special procedures and assumptions are required to relate theoretically unbiased estimates to the corresponding parameters *in vivo*, with the exception of total number. That is, provided none of the reference space is lost during tissue removal, processing, and staining, a theoretically unbiased estimate of total object number will equal the expected *in vivo* value. Number is a zero-dimensional parameter and therefore is excluded from the effects of dimensional changes such as postmortem (agonal) artifacts or changes caused by tissue processing, heating, or drying, or solvents that could cause parameter estimates to differ from *in vivo* values. For the other three first-order stereological parameters of volume, surface area, and length, the effects of nonstereological tissue processing and agonal artifacts may produce dimensional changes that will directly affect sample estimates. One can minimize the effects of variation caused by these sources by standardization of protocols and strict adherence to quality assurance procedures to ensure equivalent handling and processing of tissue.

When a source of systematic variation is suspected, every effort should be made first to identify and second to quantify the effect. For instance, you might find that tissue from group A is shrinking less than tissue from controls after an equivalent time in a formalin-based fixative. As indicated earlier, a common source of tissue shrinkage is the movement of water out of interstitial spaces as the tissue undergoes progressive formalin fixation. One possible cause of differential shrinkage could be that the treatment for group A interferes with the normal movement of water during fixation, thus leading to a systematic change in the volume of the fixed tissue. This hypothesis could be tested by snap-freezing tissue from both groups using ice-cold isopentane. Rapid freezing preserves tissue for frozen sectioning and estimation of volume without allowing time for the possible confounding effects of differential movement of water across membranes. In the early stages of a new stereology study, you should be alert to possible systematic errors related to dimensional artifacts that could complicate interpretations of sample estimates.

What Is the Source of Nonstereological Bias?

Theoretically unbiased estimates are not necessarily the outcome of theoretically unbiased methods. A theoretically unbiased method implies a theoretical concept, while a theoretically unbiased estimate is the application of the concept to a practical situation. Theoretically unbiased methods have been designed to avoid all known sources of stereological bias. However, there are a wide variety of nonstereological sources of bias that can introduce bias into estimates, despite the use of a theoretically unbiased method. Among the most common are recognition bias, ascertainment bias, tissue-processing artifacts, and operator bias. With experience, one learns to avoid these potential sources of systematic error.

Recognition Bias

Theoretically unbiased methods assume that the morphological feature of interest can be unambiguously identified. The investigator's inability to identify the feature could introduce bias into the result. In a pilot study the investigator should determine and optimize the tissue-processing protocols that will be used to quantify the object of interest. In some cases, even with clear staining, the morphological feature of interest may look similar to other features. In most cases this is a rare event and as such will not have a significant impact on the results. However, should recognition bias become more frequent, despite one's best efforts to elevate the signal of interest from the background noise in the tissue, one can keep a tally of the number of ambiguous objects, as well as how they were scored (i.e., counted or not counted). At the completion of the study, the data can be examined with and without the inclusion of these counts, and thus allow the investigator to gauge whether this source of recognition bias affects the outcome of the study. This exercise will allow the effect of the recognition bias on the outcome of the study to be determined.

Ascertainment Bias

This kind of bias occurs when estimates are based on unequal sampling probability at any stage of the sampling design. For instance, ascertainment bias occurs when a sample estimate for one population of individuals is extrapolated to a different population. In the case of human tissues, hospitals and research centers that specialize in certain diseases may receive an inordinately large sample of unusual cases involving that disease, rather than a more representative sample of typical and unusual cases. In this case, results from the unusual cases may be extrapolated to the usual disease, rather than being limited to the relatively small percentage of unusual cases from which the sample was drawn.

Tissue-Processing Artifacts

The need to process a tissue for visualization of particular features of interest introduces the possibility that nonrandom morphological changes will produce bias in the results. With experience in histology, the stereologist learns to recognize these artifacts, and, it is hoped, to avoid introducing them into tissue. Most of these artifacts are clearly recognizable; others can be minimized by careful attention to ensure equivalent processing of tissues from different groups.

Operator Bias

This form of bias arises when the person collecting the data has a preconceived idea about the outcome of a study. Operator bias is usually not a significant source of nonstereological bias for experienced investigators and those aware of the possibility of inflating or deflating estimates based on expectations. In most cases, hiding the identity of treatment groups from the person collecting the data effectively avoids the introduction of operator bias. For this reason, during data collection the identification label on slides should always be masked, even when the person collecting the data is completely unaware of the identity of the groups being compared.

How Many Animals and How Much Sampling Is Required?

Because of limited resources for research, perhaps the greatest concern among biologists and biomedical scientists is the number of individuals to be sampled and the extent to which they should be sampled. The answer to these questions depends on the biological variation of the parameter of interest and on the investigator's judgment about the biological importance of the variation observed. The optimal level of sampling is the level that captures most of the variation in the parameter within each individual, but this level of sampling is dictated by the biological variation in the parameter at the population level. With theoretically unbiased stereological methods, we can make reliable estimates of biological variation based on the observed total variation and sampling error in our estimates from a few individuals. Other than measuring it, nothing can be done about the biological variation of a parameter; this variation is set by natural selection and other forces outside the investigator's control. Sampling error, however, is controlled by the investigator, who determines the spacing between disectors, the number of sections to be analyzed throughout the reference space, the size of the geometric probes, etc. The primary consideration is that the greater the biological variation, the greater the value in sampling more individuals from the population and the less the need for high-intensity sampling within each individual. Once the mean sampling error for each group is low relative to the biological variation, further sampling within individuals will not have an appreciable effect on the pooled variance of the sample estimate. At that point, the investigator can only sample more individuals to achieve further reductions in the pooled variance (see Chapter 10).

What If I Cannot Define the Boundaries of My Reference Space?

Stereology requires that one be able to precisely define the boundaries of a reference space; or for populations of objects, that one be able to unambiguously identify when the object is sampled. In theory, an estimate relates to a well-defined, bounded structure. Mathematically, the reference space and the population of objects are deterministic factors in the estimate. If a reference space is defined without much reproducibility, say, one day as relatively large, the next day as small, the corresponding estimates of these spaces (e.g., total volume, total number of cells) will vary accordingly. A distinct description of the anatomical and/or functional borders is necessary to communicate the boundaries of a reference space in reporting results.

   As described in Chapter 10, an anatomically well-defined reference space or identifiable population of objects permits the user to partition the total observed variation into two sources: biological variation and sampling error. With this information we can optimize the sampling design for maximum efficiency by keeping the sampling error low relative to the biological variation. A poorly defined reference space or population of cells may confound partitioning of the total variation by including unaccounted-for variation arising from the delineation of the reference space, rather than biological variation or sampling error. If this delineation-related variation is systematic, increasing the sampling intensity or sampling more individuals will not reduce this variation; it simply becomes incorporated into the sample estimate as bias.

Is a Reference Space the Same as a Region of Interest?

A reference space is not the same as a region of interest although the terms are frequently interchanged. The term *region of interest* is used in the analysis of digital images (image analysis). The first step of image analysis is to digitize the image, that is, convert from a high-resolution analog image to a relatively low-resolution digital image. The loss of resolution during the analog-to-digital conversion occurs when the analog image is divided into its picture elements (pixels) with a range of 256 gray levels of contrast. In the second step of image analysis, the operator drags a mouse or pointer over the image to select an ROI, a population of pixels for further analysis. The analysis may include color enhancement, high/low pass filters, or segmentation according to threshold gray levels selected by the operator. The ROI feature allows the operator to indicate the specific set of pixels within the image to be analyzed. Therefore, the main consideration when delineating an ROI in image analysis is to identify pixels for analysis, rather than to specify the anatomical boundaries of a region for subsequent systematic-random sampling.

For regions of tissue, distinct anatomical boundaries are the critical determinant of a reference space. For populations of objects, the critical factor is that the operator be able to determine whether a particular object (cell, synapse, etc.) can be identified as part of the target population. For the estimation of stereological parameters using theoretically unbiased methods, these boundaries determine the magnitude of the estimate. A sample estimate based on a poorly defined reference space or poorly defined population of objects or features may vary in a systematic manner and thus include an unmeasurable quantity of variation unrelated to biological variation or sampling error.

An ROI can be a useful reference space, provided its anatomical boundaries are defined according to a reproducible strategy. That is, an ROI can be a well-defined reference space if it is an anatomically distinct structure with boundaries that can be easily and reproducibly defined, both by the same individual at different times (intra-rater reliability) and by different individuals working independently (inter-rater reliability).

An easy way to test the definition of any reference space is to have one or more persons indicate the borders of the space at low magnification. This can be done by marking the boundaries on the back of a glass slide, on a photograph, or on a saved digital image. These initial outlines can then be compared with a repeat delineation of the same structure by the same operator or a different operator to measure intra- and inter-operator error, respectively. The outlined areas can be quantified using a point grid or a pixel-counting program. If the same operator or two operators working independently consistently outline quantifiably similar areas, then they are most likely using a similar strategy arising from features in the tissue, rather than the operator's imagination. The boundaries of an ideal reference space will vary little from one operator to the next and from one session to the next.

Are Theoretically Unbiased Methods Always Based on Theoretically Unbiased Principles?

Theoretically unbiased stereological methods are based on proven principles of theoretically unbiased sampling and estimation. The methods consist of a theoretically unbiased estimator and an unbiased geometric probe. Using systematic-random sampling, the probe is placed at random in the tissue and throughout the entire reference space. The number of intersections between the probe and the objects of interest is proportional to the expected value of the parameter of interest. Provided nonstereological bias does not introduce systematic error, the results of a theoretically unbiased method will be theoretically unbiased; however, avoidance of nonstereological bias is dependent on the investigator rather than the sampling approach or geometric probe.

How Does Modern Stereology Differ from Conventional Morphometry?

For decades conventional morphometry in the biological sciences was heavily influenced by classical geometry (e.g., "Assume a cell is a sphere . . ."). However, when applied to populations of biological structures that do not fit the assumptions of these models, conventional morphological approaches invariably introduce bias (systematic error). Like the error associated with poorly defined reference spaces and populations of objects, the systematic error introduced by faulty assumptions and models cannot be quantified or removed; rather, it becomes a permanent part of the sample estimate. In contrast, modern stereology is based on stochastic geometry and probability theory, and specifically avoids using morphometric methods that rely on assumptions, models, and correction factors.

Has a Parameter Ever Been Estimated by Theoretically Unbiased Stereology and Conventional Methods?

Several parameters have been estimated using both theoretically unbiased methods and conventional approaches. In particular, image analysis is a conventional morphometry method that in theory provides the potential for relatively rapid speed, with the same level of accuracy as theoretically unbiased stereology. In 1981 Gundersen et al. quantified the profile areas of cells, nuclei, and nucleoli by a conventional (pixel counting) method and by a theoretically unbiased stereological technique (point counting). The results of this study showed that both methods could give results that were similar with regard to accuracy. However, the two approaches differed markedly with respect to precision and efficiency. The pixel-counting approach showed markedly lower levels of precision (higher variation), unless time and effort were used to carefully outline the structures of interest and to edit data from extraneous structures out of the estimate. The authors suggest that this discrepancy arises from the sensorimotor processing required to collect data for each method. The pixel-counting approach requires a "brain–eye–hand" combination for outlining structures and editing the image. In contrast, the point-counting method only requires a "brain–eye" combination to count points inside the reference space. That is, movements of the hand do not introduce errors when point counting.

Will Computers Someday Take Over the Job of Recognizing Biological Objects?

Biological objects are naturally occurring and hence show high morphological variability on tissue sections, in contrast to classical objects and models used as the mathematical basis for computer software. In theory, once computer software can "learn" to recognize biological morphology with the same level of accuracy as a trained human, computerized approaches will begin to take over the role currently played by humans sitting before computers. However, at the dawn of the twenty-first century, automatic stereology using computer programs cannot generate theoretically unbiased estimates without introducing substantial amounts of systematic error. Perhaps someday computerized systems will be able to discriminate subtle changes in morphology, color, and other features currently recognized only by the eyes of trained and experienced humans.

Unbiased Stereology Quantification System
System Schematic

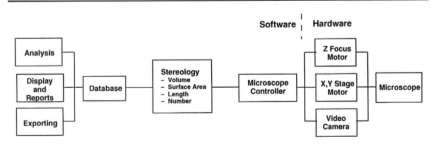

Figure 12-1. Diagram showing the major components and processes underlying integrated hardware–software systems for computerized stereology analysis

This is not to say that computer systems cannot facilitate unbiased stereology studies. Figure 12-1 shows the design of a computerized stereology system. These types of hardware–software systems take over most of the time- and labor-consuming tasks of stereological analysis, focusing the operator's efforts on precise and accurate recognition of objects and features of interest in tissue sections.

Appendix

Conceptual Framework of Modern Stereology

A *well-defined* historical and conceptual framework exists for the application of new stereology to biological tissue.

1. Developed by materials scientists, mathematicians, and biologists since early 1960s.

2. Can be applied to all well-defined populations of biological structures that can be unambiguously identified.

3. Tissue-processing requirements differ from those of assumption-based methods.

4. Avoids all known sources of methodological biases (e.g., "assume a cell is a sphere").

5. Avoids inappropriate correction formulas.

6. Focuses on total parameters (e.g., total cell number), not ratios or densities (e.g., cell number per unit).

7. Estimates first-order parameters of biological structures (e.g., volume, surface area, length, number) and their variability using a small sample of the population of interest.

8. Uses highly efficient systematic sampling.

9. Efficiency is based on biological variability.

10. Estimates refer to a biologically meaningful reference space in a defined population.

11. Sampling can be optimized for maximum efficiency ("Do more less well").

12. Allows optimization for maximum efficiency.

13. Avoids tissue-processing artifacts, for example, shrinkage or expansion, lost caps.

14. Strong mathematical foundation in stochastic geometry and probability theory, but . . .

15. Advanced mathematical background not required for users.

16. Does not require computerized hardware–software systems, but . . .

17. Computerized stereology systems are very efficient.

18. Statistical power for studies of the same parameter across populations is cumulative.

19. Increasingly preferred by journal editors and grant review study groups as state-of-the-art approach for morphometric analysis of biological structures.

20. Worldwide cooperation possible, in theory, through web-accessible databases.

Glossary

Accuracy. The approach to the expected value of a data point.

Anisotropy. A preferred spatial orientation.

Arbitrarily shaped. Variable in 3-D; nonclassical.

Area. First-order, 2-D stereological parameter.

Area fraction. Structural area relative to a reference area; equivalent to volume fraction according to Delesse principle

Area sampling fraction (asf). Fraction of total area sampled.

Ascertainment bias. Systematic error in sampling individuals from a target population.

Assumption-based morphometry. Morphometry based on nonverifiable assumptions or models.

Between-section variation. Random variation caused by variability between adjacent sections.

Bias. Error in a systematic direction.

Biological variation (error). Variability in a parameter caused by individual differences.

Biological variance. Proportion of observed variance caused by individual differences.

Cardinality. The number of elements in a defined set.

Classical geometry. The field of geometry that deals with classically shaped structures; model based.

Coefficient of error (CE). Second-order stereological parameter for within-sample variation (sampling error).

Coefficient of variation (CV). Second-order stereological parameter for total observed variation.

Correction factor. Assumption-based component of a morphometric equation.

Cycloid. A sine-weighted line probe for estimation of surface area and length from vertical sections and vertical slices.

Density. A ratio estimator involving a first-order stereological parameter (numerator) and its reference volume (denominator), e.g., length density (L_v), numerical density (N_v).

Design-based stereology. Morphometric approach to estimating structural parameters in fixed, defined populations using sampling and probes that conform to randomness, as required by probability and stochastic geometric theory.

Dependent (systematic, nonrandom) sampling. Scheme in which adjacent structures are sampled in serial order.

Digitization. Digital-to-analog process that converts a light image to picture elements (pixels).

Disector. A three-dimensional sampling and counting probe.

Distribution. The spread of values across a target population.

Do more less well. Optimal strategy for reducing observed variance in structural parameters by sampling less within each individual in favor of sampling a greater number of individuals.

Efficiency. Rate at which precision increases as a function of time.

Embedding. Placement of tissue in a hard matrix for sectioning.

Error. Variation in a parameter.

Error analysis. Post hoc process in which observed error is partitioned into sources.

Error variance. Random error in a population; noise.

Estimate. The approximate calculation of a parameter.

Exclusion lines. Lower and left lines on a theoretically unbiased counting frame; exclusion planes in 3-D.

Expected value. The true or expected value of a parameter for the population.

Fixation. Stabilizing of tissue proteins prior to analysis.

Fractionator. Sampling procedure in which total parameters are estimated from measurements in a known fraction of the total reference space.

Inclusion lines. Upper and right lines on a theoretically unbiased counting frame; inclusion planes in 3-D.

Individual differences. Variability between individual members of a sample or population.

Interpretation bias. Systematic error in an estimate caused by the experimenter favoring a particular outcome a priori.

Isotropy. Possessing properties that have the same values when measured along axes in all directions.

Isotropic-uniform-random (IUR). Method for sampling a structure in which all orientations in three dimensions have equal weight.

Length. First-order stereological parameter for lineal (1-D) structures; estimated without bias using a 2-D probe (surface).

Line probe. Grid of uniformly oriented test lines.

Mean (average). The central tendency of a normal distribution.

Mean object volume (MOV). A parameter of average object size across a target population.

Method error. Random within-sample variation.

Microtome. Device for sectioning tissue.

Morphometry. The methodology used to measure structure; quantitative morphology.

Nonstereological bias. Systematic variation caused by practical applications of unbiased methods to biological tissue; uncertainty.

Nugget effect. Variability in an estimate caused by the smallest part of a structure that contributes to the observed variation in a parameter; within-section variation; stereological noise.

Number. First-order stereological parameter for 0-D structures; estimated without bias using a 3-D probe.

Number-weighted. Sampling approach in which all values in a defined number distribution have an equal probability of being sampled.

Object. A structure identified at high magnification.

Orientation bias. Systematic variation caused by a preferred three-dimensional orientation.

Optical disector. A theoretically unbiased method for estimating density using a stack of optical sections to sample the topmost points of three-dimensional objects in a known volume of a defined reference space.

Optical fractionator. A theoretically unbiased method for estimating total number; uses a stack of optical sections to determine the number of objects in a known fraction of a defined reference space.

Optimization. Post hoc strategy using error analysis to direct sampling effort at the primary sources of variability in estimates.

Overprojection. Excess of information in an image e.g., excess thickness of transparent embedding material relative to an opaque object of interest.

Pappas–Guldinius theorem. Mathematical tenet which holds that the mean object volume for a defined population can be estimated without bias from average lengths of isotropic lines radiating from a fixed point to the border on random profiles through the objects.

Parameter. A measured quantity in a defined population.

Pattern recognition. Technique of computerized image analysis for automatic identification of structures based on defined characteristics.

Paraffin embedding. A processing technique that uses paraffin to physically stabilize tissue for sectioning; allows use of histochemistry and immunocytochemistry for stereology of objects.

Pilot study. Preliminary analysis of a small number of samples to obtain rough data and optimize study parameters.

Plastic embedding. A processing technique using plastic to harden a tissue block for sectioning.

Point counting. Probability-based technique for theoretically unbiased estimation of area.

Population. A collection of objects that have at least one attribute in common, from which random samples are drawn for stereological estimates.

Postmortem artifact. Tissue changes that begin at the death of an organism or at separation of tissue from its blood supply.

Precision. Extent to which data points cluster around a central tendency.

Probe. A zero-to-three-dimensional test system of points, lines, planes, or volumes used for sampling and estimating structural parameters. To be theoretically unbiased, probes must be positioned at random with respect to the structure of interest; for surface and length estimates, probe–structure intersections must be isotropic.

Probe sampling fraction (psf). In the formula to estimate length and length density using isotropic sphere probes, the ratio of sampling box volume to sphere surface area.

Profile. Cross-section through an object that is sampled by a two-dimension plane (e.g., knife blade).

Quadratic approximation formula. Equation used to estimate within-sample variation (sampling error) for dependent sampling.

Random sampling. Sampling process in which each sample is independent of every other sample.

Recognition bias. Systematic error caused by the inability of the experimenter to consistently identify structures of interest.

Reference space. Biologically important region unambiguously defined by natural borders.

Reference trap. Systematic error arising from the use of density estimators to measure absolute quantities; changes in size of the reference space can affect density without changing first-order stereological parameters of the structure of interest.

Region. The level of structure identified at high magnification.

Region of interest. In image analysis, the boundaries of the structure to be analyzed.

Rotator. An application of the Pappas–Guldinius theorem for efficient estimation of mean object volume on random profiles; uses isotropic lines from a fixed point in the profile to the edge of the profile.

Sample. A subset of a population (noun); to obtain a subset of a population (verb).

Sampling error (within-sample error). Random variation related to the level of sampling within an individual.

Sampling interval (k). The period between successive samples.

Section sampling fraction (ssf). The number of sections analyzed divided by the total number of sections through the reference space.

Shape bias. Variation of data in a systematic manner caused by variability in the shape of objects.

Size bias. Variation of data in a systematic manner caused by variability in the size of objects.

Snap-freezing. A procedure that uses liquid nitrogen to rapidly freeze tissue; retains water and ions within cells and prevents shrinkage or expansion of the reference space.

Space ball. An isotropic sphere probe for estimating length and length density.

Standard error of the mean (SEM). The standard deviation divided by the square root of the number of individuals in the sample.

Stereological bias. Systematic variation caused by theoretical assumptions, models, correction factors.

Stereology. The analysis of three-dimensional structures (Greek).

Stereologer. An integrated hardware-software system for theoretically unbiased stereological analysis of biological structures.

Stochastic geometry. Methodology for mathematical analysis of arbitrarily shaped objects.

Systematic-uniform-random (SUR). Sampling approach that allows all parts of a reference space to be sampled in an efficient manner and without methodological bias.

Systematic error. Nonrandom variation that causes sample data to deviate from expected values.

Theoretically unbiased method. An approach that avoids systematic error; with increased sampling, sample estimates using a theoretically unbiased method approach their expected value for the population parameter.

Theoretically unbiased counting frame. An areal frame that ensures that the number distribution of objects is sampled with equal probability.

Thickness sampling fraction (tsf). The percentage of section thickness occupied by the height of a disector.

Underprojection. Lack of information on projected images caused by embedding media that are opaque relative to the object of interest.

Variation (variability). Error; may be systematic (bias) or nonsystematic (random).

Vertical section. A random plane parallel to the axis of rotation.

Vertical axis. The axis of rotation.

Vertical-uniform-random (VUR). In conjunction with cycloid line probes, sampling method to avoid systematic error (orientation bias) when estimating surface area and length of anisotropic structures.

Vibratome. A sectioning apparatus that uses a vibrating knife blade to cut sections through tissue.

Virtual sectioning. Optical sectioning technique that uses a stack of parallel focal planes to probe objects.

Volume. A three-dimensional, first-order stereological parameter.

Volume-weighted sample. A subset of objects or regions sampled on the basis of size.

Within-sample error. Random variation arising from sampling within an individual; measured by CE.

Within-section variation. Random variation within a section caused by the smallest part (nugget) of a region or object; the nugget effect (stereological noise).

Bibliography

Abercrombie, M. (1946). Estimation of nuclear population from microtome sections. *Anatomical Record, 94:* 239–247.

Baddeley, A. J., Gundersen, H. J. G., and Cruz-Orive, L. M. (1986). Estimation of surface area from vertical sections. *Journal of Microscopy, 142:* 259–276.

Bradbury, S. (1983). Commercial image analysers and the characterizations of microscopical images. *Journal of Microscopy, 131:* 203–210.

Brody, H. (1955). Organization of the cerebral cortex III. A study of aging in the human cerelarial cortex. *Journal of Comparative Neurology, 102:* 511–556.

Buffon, G.-L. Leclerc, Comte de (1777). Essai d'arithmétique morale. Supplement à *l'Historie Naturelle* Vol. 4, Imprimerie Royale, Paris.

Calhoun, M. E., and Mouton, P. R. (2001). Length measurement: New developments in neurostereology and 3D imagery. *Journal of Chemical Neuroanatomy, 21:* 257–265.

Calhoun, M., Jucker, M., Martin, L., Thinakaren, G., Price, D., and Mouton, P. R. (1996). Comparative evaluation of synaptophysin-based methods for quantification of synapses. *Journal of Neurocytology, 25:* 821–828.

Calhoun, M. E., Kurth, D., Phinney, A. L., Long, J. M, Hengemihle, J., Mouton, P. R., Ingram, D. K., and Jucker, M. (1998). Hippocampal neuron and synaptophysin-positive bouton number in aging C57BL/6 mice. *Neurobiology of Aging, 19:* 599–606.

Casey-Smith, J. R. (1988). Expressing stereological results 'per cm^3' is not enough. *Journal of Pathology, 156:* 263–265.

Cavalieri, B. (1635). *Geometria Indivisibilibus Continuorum.* Typis Clementis Ferronij. Bononi. Reprinted (1966) as *Geometria degli Indivisibili.* Unione Tipografico-Editrice Torinese, Torino.

Coggeshall, R. E., and Lekan, H. A. (1996). Methods for determining number of cells and synapses: A case for more uniform standards of review. *Journal of Comparative Neurology, 364:* 6–15.

Cruz-Orive, L. M. (1989). On the precision of sampling: A review of Matheron's transitive methods. *Journal of Microscopy, 153:* 315–333.

Cruz-Orive, L. M. (1989). Second-order stereology: Estimation of second-moment volume measures. *Acta Stereologica, 8* (2): 641–646.

Cruz-Orive, L. M. (1993). Systematic sampling in stereology. *Bulletin of the International Statistical Institute, 55* (2): 451–468.

Cruz-Orive, L. M. (1994). Toward a more objective biology. *Neurobiology of Aging, 15* (3): 377–378.

Cruz-Orive, L. M. (1997). Stereology of single objects. *Journal of Microscopy, 183:* 93–107.

Cruz-Orive, L. M., and Weibel, E. R. (1981). Sampling designs for stereology. *Journal of Microscopy, 122:* 235–257.

Cruz-Orive, L. M., and Weibel, E. R. (1990). Recent stereological methods for cell biology: A brief survey. *American Journal of Physiology, 258 (Lung Cell Mol. Physiol. 2):* L148- L156.

Dam, A. M (1979). Brain shrinkage during histological procedures. *Journal für Hirnforschung, 20:*115–119.

de Groot, D. M. G., and Bierman, P. B. (1986). A critical evaluation for estimating the number and density of synapses. *Journal of Neuroscience Methods, 18:* 79–101.

DeHoff, R. T., and Rhines, F. N. (1961). Determination of number of particles per unit volume from measurements made on random plane sections: The general cylinder and the ellipsoid. *Transactions Metallurgical Society, AIME, 221:* 975–982.

Delesse, M. A. (1847). Procédé mécanique pour déterminer la composition des roche. *Comptes Rendus de l'Academie des Sciences, Paris, 25:* 544–545.

Double, K. L., Halliday, G. M., Kril, J. J., et al. (1996). Topography of brain atrophy during normal aging and Alzheimer's disease. *Neurobiology of Aging, 17:* 513–521.

Elias, H. (1971). Three-dimensional structure identified from single sections. *Science, 174:* 993–1000.

Elias, H., and Schwartz, D. (1969). Surface areas of the cerebral cortex of mammals determined by stereological methods. *Science, 166:* 111–113.

Elias, H., Henning, A., and Schwartz, D. (1971). Stereology: Applications to biomedical research. *Physiological Reviews, 1:* 158–200.

Elias, H., Pauly, J. E., and Burns, E. R. (1978). *Histology and Human Microanatomy,* 4th ed. Wiley, New York.

Floderus, S. (1944). Untersuchungenü ber den Bau der menschlichen Hypophyse mit besonderer Berücksichtigung der quantitativen mikromorphologischen Verhältnisse. *APMIS 53:* 1–176.

Flood, D. G., and Coleman, P. D. (1988). Neuron numbers and sizes in aging brain: Comparisons of human, monkey, and rodent data. *Neurobiology of Aging, 9:* 453–463.

Geinisman, Y., Gundersen, H. J. G., Van Der Zee, E., and West, M. J. (1996). Stereological estimation of the total number of synapses in a brain region. *Journal of Neurocytology, 25:* 805–819.

Geinisman, Y., Gundersen, H. J. G., Van Der Zee, E., and West, M. J. (1997). Towards obtaining estimates of the total number of synapses in a brain region: Problems of primary and secondary importance. *Journal of Neurocytology, 26:* 711–713.

Glagolev, A. A. (1933). On the geometrical methods of quantitative mineralogic analysis of rocks. *Transactions of the Institute of Economic Mineralogy and Metallurgy, 59:* 1.

Glenny, R. W., and Robertson, H. T. (1991). Fractal modeling of pulmonary blood flow heterogeneity. *Journal of Applied Physiology, 70:* 1024–1030.

Gokhale, A. M. (1990). Estimation of curve length in 3-D using vertical slices. *Journal of Microscopy 159:* 133–141.

Gokhale, A. M. (1992). Estimation of length density L_v, from vertical slices of unknown thickness. *Journal of Microscopy, 167:* 1–8.

Gokhale, A. M. (1993). Utility of the horizontal slice for stereological characterization of lineal features. *Journal of Microscopy, 170:* 3–8.

Gundersen, H. J. G. (1977). Notes on the estimation of the numerical density of arbitrary profiles: The edge effect. *Journal of Microscopy, 143:* 3–45.

Gundersen, H. J. G. (1979). Estimation of tubule or cylinder LV, SN, and VV on thick sections. *Journal of Microscopy, 117:* 333–345.

Gundersen, H. J. G. (1986). Stereology of arbitrary particles. A review of number and size estimators and the presentation of some new ones, in memory of William R. Thompson. *Journal of Microscopy 143:* 3–45.

Gundersen, H. J. G. (1988). The nucleator. *Journal of Microscopy, 151:* 3–21.

Gundersen, H. J. G., and Jensen, E. B. (1983). Particle sizes and their distributions estimated from line- and point-sampled intercepts, including graphical unfolding. *Journal of Microscopy, 131:* 291–310.

Gundersen, H. J. G., and Jensen, E. B. (1985). Stereological estimation of the volume-weighted mean volume of arbitrary particles observed on random sections. *Journal of Microscopy, 138:* 127–142.

Gundersen, H. J. G., and Jensen, E. B. (1987). The efficiency of systematic sampling in stereology and its prediction. *Journal of Microscopy, 147:* 229–263.

Gundersen, H. J. G., and Østerby, R. (1981). Optimizing sampling efficiency of stereological studies in biology: Or "Do more less well!" *Journal of Microscopy, 121:* 65–73.

Gundersen, H. J. G., Boysen, M., and Reith, A. (1981). Comparison of semiautomatic digitizer-table and simple point counting performance in morphometry. *Virchows Archives, 37:* 3–45.

Gundersen, H. J. G., Bagger, P., Bendtsen, T. F., Evans, S. M., Korbo, L., Marcussen, N., Møller, A. M., Nielsen, K., Nyengaard, J. R., Pakkenberg, B., Sorensen, F., Vesterby, A., and West, M. J. (1988). The new stereological tool: Disector, fractionator, nucleator and point sampled intercepts and their use in pathological research and diagnosis. *APMIS 96:* 857–881.

Gundersen, H. J. G., Bendtsen, T. F., Korbo, L., Marcussen, N., Evans, S. M., Møller, A. M., Nielsen, K., Nyengaard, J. R., Pakkenberg, B., Sorensen, F., Vesterby, A., and West, M. J. (1988). Some new, simple and efficient stereological methods and their use in pathological research and diagnosis. *APMIS 96:* 379–394.

Gundersen, H. J. G., Boyce, R. W., Nyengaard, J. R., and Odgaard, A. (1993). The conneulor: Estimation of connectivity using physical disectors under projection. *Bone, 14:* 217–222.

Gundersen, H. J. G., Jensen, E. B., Kiêu, K., and Nielsen, J. (1999). The efficiency of systematic sampling in stereology—Reconsidered. *Journal of Microscopy, 193:* 199–211.

Gupta, M., Mayhew, M., Bedi, K. S., Sharma, A. K., and White, F. H. (1983). Inter-animal variation and its influence on the overall precision of morphometric estimates based on nested sampling designs. *Journal of Microscopy, 131:* 147–153.

Haug, H. (1986). History of neuromorphometry, *Journal of Neuroscience Methods, 18:* 1–17.

Haug, H., Kühl, S., Mecke, E., Sass, N., and Wasner, K. (1984). The significance of morphometric procedures in the investigation of age changes in cytoarchitectonic structures of human brain. *Journal für Hirnforschung, 25:* 353–374.

Hanstede, J. G., and Gerrits, P. O. (1983). The effects of embedding in water-soluble plastics on the final dimensions of liver sections. *Journal of Microscopy, 131:* 79–86.

Hedreen, J. C. (1998). What was wrong with the Abercrombie and empirical cell counting methods? A review. *Anatomical Record, 250* (3): 373–380.

Hedreen, J. C. (1998). Lost caps in histological counting methods. *Anatomical Record, 250* (3): 366–372.

Hedreen, J. C. (1999). Unbiased stereology? *Trends in Neuroscience, 22:* 346.

Henery, C. C., and Mayhew, T. (1989). The cerebrum and cerebellum of the fixed human brain: Efficient estimates of volumes and cortical surface areas. *Journal of Anatomy, 167:* 167–180.

Hennig, A. (1963). Length of a three-dimensional linear tract. *Proceedings 1st International Congress on Stereology, 44:* 1–8.

Howard, C. V., Cruz-Orive, L. M., and Yaegashi, H. (1992). Estimating neuron dendritic length in 3D from total vertical projections and from vertical slices. *Acta Neurologica Scandinavica,* Supp. *137:* 14–19.

Jarvis, L. R. (1988). Microcomputer video image analysis. *Journal of Microscopy, 150:* 83–97.

Jensen Vedel, E. B. (1987). Design- and model-based stereological analysis of arbitrarily shaped particles. *Scandinavian Journal of Statistics, 14:* 161–180.

Jensen Vedel, E. B. (1991). Recent developments in the stereological analysis of particles. *Annals of the Institute of Statistical Math, 43:* 455–468.

Jensen Vedel, E. B., and Gundersen, H. J. G. (1989). Fundamental stereological formulae based on isotropically oriented probes through fixed points with applications to particle analysis. *Journal of Microscopy, 153:* 249–267.

Jensen Vedel, E. B., and Gundersen, H. J. G. (1992). The Rotator. Research Report no. 247, Department of Theoretical Statistics, Institute of Mathematics and Stereological Research Laboratory, University of Aarhus, Denmark.

Jensen Vedel, E. B., and Gundersen, H. J. G. (1993). The Rotator. *Journal of Microscopy, 170:* 35–44.

Jensen Vedel, E. B., and Sørenson, F. B. (1991). A note on stereological estimation of the volume-weighted second moment of particle volume. *Journal of Microscopy, 164:* 21–27.

Jensen Vedel, E. B., Baddley, A. J., Gundersen, H. J. G., and Sundberg, R. (1985). Recent trends in stereology. *International Statistics, 53:* 99–108.

Jensen Vedel, E. B., Kiêu, K., and Gundersen, H. J. G. (1990). Second-order stereology. *Acta Stereologica, 9:* 15–35.

Jones, D. G., Itarat, W., and Calverley, R. K. S. (1992). Perforated synapses and plasticity: A developmental overview. *Molecular Neurobiology, 5:* 217–228.

Larsen, O. J., Gundersen, H. J. G., and Nielsen, J. (1998). Global spatial sampling with isotropic virtual planes: Estimators of length density and total length in thick, arbitrarily orientated section. *Journal of Microscopy, 191:* 238–248.

Long, J. M., Ingram, D. K., Kalehua, A., and Mouton, P. R. (1998). Stereological estimation of microglia number in hippocampal formation of the mouse brain. *Journal of Neuroscience Methods, 84:* 101–108.

Long, J. M., Kalehua, A., Muth, N. J., Calhoun, M. E., Jucker, M., Hengemihle, J. M., Ingram, D. K., and Mouton, P. R. (1998). Stereological analysis of astrocyte and microglia in aging mouse hippocampus. *Neurobiology of Aging, 19:* 495–503.

Long, J. M., Mouton, P. R., Jucker, M., and Ingram, D. K. (1999). What counts in brain aging? Design-based stereology analysis of cell number. *Journal of Gerontology, 54:* B407–417.

Loud, A. V. (1968). A quantitative stereology description of the ultrastructure of normal rat liver parenchymal cells. *Journal of Cellular Biology, 37:* 27–46.

Mandelbrot, B. B. (1967). How long is the coastline of Great Britain? Statistical self-similarity and fractal dimension. *Science 155:* 636–638.

Mandelbrot, B. B. (1977). *Form, Chance, and Dimension.* Freeman, New York.

Mandelbrot, B. B. (1983). *The Fractal Geometry of Nature.* Freeman, New York.

Mandelbrot, B. B. (1999). A multifractal walk down Wall Street. *Scientific American,* February, 70–73.

Matheron, G. (1972). Random set theory and applications to stereology, *Journal of Microscopy, 95:* 15–23.

Mathieu, O., Cruz-Orive, L. M., Hoppeler, H., and Weibel, E. R. (1981). Measuring error and sampling variation in stereology: Comparison of the efficiency of various methods for planar image analysis. *Journal of Microscopy, 121:* 75–88.

Mattfeldt, T., Mall, G., von Herbay, A., and Moller, P. (1989). Stereological investigation of anisotropic structure with the orientator. *Acta Stereologica, 8:* 671–677.

Mayhew, T. M. (1992). A review of recent advances in stereology for quantifying neural structure. *Journal of Neurocytology, 21:* 313–328.

Mayhew, T. M., and Olsen, D. R. (1991). Magnetic resonance imaging (MRI) and model-free estimates of brain volume determined using the Cavalieri principle. *Journal of Anatomy, 178:* 133–144.

Michel, R. P., and Cruz-Orive, L. M. (1988). Application of the Cavalieri principle and vertical sections method to lung: Estimation of volume and pleural surface area. *Journal of Microscopy, 150:* 117–136.

Miles, R. (1976). Precise and general conditions for the validity of a comprehensive set of stereological formulas. *Journal of Microscopy, 107:* 211–220.

Møller, A., Strange, P., and Gundersen, H. J. G. (1990). Efficient estimation of cell volume and number using the nucleator and the disector. *Journal of Microscopy, 159:* 61–71.

Mouton, P. R., Pakkenberg, B., and Gundersen, H. J. G., and Price, D. L. (1994). Absolute numbers and size of pigmented locus coeruleus neurons in the brains of young and aged individuals. *Journal of Chemical Neuroanatomy 7:* 185–190.

Mouton, P. R., Price, D. L., Walker, L. C. (1997). Empirical assessment of total synapse number in primate neocortex. *Journal of Neuroscience Methods, 75:* 121–128.

Mouton, P. R., Martin, L. J., Calhoun, M. E., Dal Forno, G., and Price, D. L. (1998). Cognitive decline strongly correlates with cortical atrophy in Alzheimer's dementia. *Neurobiology of Aging, 19:* 371–377.

Mouton, P. R., Gokhale, A. M., Ward, N., and West, M. J. (2002). Stereological length estimation using spherical probes. *Journal of Microscopy,* in press.

Nielsen, K., Berild, G. H., Brun, E., et al. (1989). Stereological estimation of mean nuclear volume in prostatic cancer, the reproducibility and the possible value of estimations on repeated biopsies in the course of disease. *Journal of Microscopy, 154:* 63.

Nyengaard, J. R., and Gundersen, H. J. G. (1992). The isector: A simple and direct method for generating isotropic, uniform random sections from small specimens. *Journal of Microscopy, 150:* 1–20.

Ohm, T. G., Busch, C., and Bohl, J. (1997). Estimation of neuronal numbers in the human nucleus coeruleus during aging. *Neurobiology of Aging, 18:* (4), 393–399.

Pache, J. C., Roberts, N., Zimmerman, A., Vock, P., and Cruz-Orive, L. M. (1993). Vertical LM sectioning and parallel CT scanning designs for stereology: Applications to human lung. *Journal of Microscopy, 170:* 3–24.

Pakkenberg, B., and Gundersen, H. J. G. (1988). Total number of neurons and glial cells in human brain nuclei estimated by the disector and the fractionator. *Journal of Microscopy, 150:* 1–22.

Pakkenberg, B., and Gundersen, H. J. G. (1997). Neocritical neuron number in humans: Effect of sex and age. *Journal of Comparative Neurology 384:* 312–320.

Pakkenberg, B., Bosen, J., Albeck, M., and Gjerris, F. (1989). Efficient estimation of total ventricular volume from CT-scans by a stereological method. *Neuroradiology 31:* 413–427.

Pakkenberg, B., Andersen, B. B., Jensen, G. B., Korbo, L., Mouton, P. R., Møller, A., Regeur, L., and Øster, S. (1992). The impact of the new stereology on the neurosciences-neurostereology. *Acta Stereologica 11,* Supp. (1): 157–164.

Paumgartner, D., Losa, G., and Weibel, E. R. (1981). Resolution effect on the stereological estimation of surface and volume and its interception in terms of fractal dimension. *Journal of Microscopy, 121:* 51–63.

Regeur, L., and Pakkenberg, B. (1989). Optimizing sampling designs for volume measurements of components of human brain using a stereological method. *Journal of Microscopy, 155:* 113–121.

Roberts, N., Cruz-Orive, L. M., Reid, N. M. K., Brodie, D. A., Bourne, M., and Edwards, R. H. T. (1993). Unbiased estimation of the human body composition by the Cavalieri method using magnetic resonance imaging. *Journal of Microscopy, 171:* 239–253.

Roberts, N., Garden, A. S., Cruz-Orive, L. M., Whitehouse, G. H., and Edwards, R. H. T.

(1994). Estimation of fetal volume by magnetic resonance imaging and stereology. *British Journal of Radiology, 67* (803): 1067–1077.

Rosival, A. (1898). Über geometrische Gesteinsanalysen. *Verh. K. K. Geol. Reichsanst., Wien,* 143–175.

Saper, C. B. (1996). Any way you cut it: A new journal policy for the use of counting methods. *Journal of Comparative Neurology, 354:* 5.

Saper, C. B. (1997). Counting on our reviewers to set the standards. *Journal of Comparative Neurology, 386:* 1.

Schmitz, C., Korr, H., and Heinsen, H. (1999). Design-based counting techniques: The real problems. *Trends in Neuroscience, 22:* 345.

Smith, C. S., and Gutteman, L. (1953). Measurement of internal boundaries in three-dimensional structures by random sectioning. *Transactions American Institute Mining, Metallurgy, and Petroleum Engineers, 197:* 81–92.

Sørensen, F. B. (1992). Quantitative analysis of nuclear size for objective malignancy grading: A review with emphasis on new, unbiased stereologic methods. *Biology of Disease, 66* (1): 4–23.

Sterio, D. C. (1984). The unbiased estimation of number and sizes of arbitrary particles using the disector. *Journal of Microscopy, 134:* 127–136.

Stocks, E. A., McArthur, J., Griffin, J. A., and Mouton, P. R. (1996). An unbiased method for estimation of total nerve fiber length. *Journal of Neurocytology, 25:* 11–18.

Subbiah, P., Mouton, P. R., Fedor, H., McArthur, J., and Glass, J. (1996). Stereological analysis of cerebral atrophy in human immunodeficiency virus-associated dementia. *Journal of Neuropathology and Experimental Neurology, 55:* 1032–1037.

Tandrup, T. (1993). A method for unbiased and efficient estimation of number and mean volume of specified neuron subtypes in rat dorsal root ganglion. *Journal of Comparative Neurology, 329:* 269–276.

Thomson, W. R. (1930). Quantitative microscopic analysis. *Journal of Geology, 38:* 193.

Thomson, W. R. (1932). The geometric properties of microscopic configurations. I. General aspects of projectometry. *Biometrika, 24:* 21–26.

Vesterby, A., Gundersen, H. J. G., and Melsen, F. (1989). Star volume of marrow space and trabeculae of the first lumbar vertebra: Sampling efficiency and biological variation. *Bone, 10:* 7–13.

Von Bonin, G. (1973). About quantitative studies on the cerebral cortex. *Journal of Microscopy, 99:* 75–83.

Weibel, E. R. (1979). *Stereological Methods,* vol. 1., *Practical Methods for Biological Morphometry,* Academic Press, London.

Weibel, E. R. (1989). Measuring through the microscope: Development and evolution of stereological methods. *Journal of Microscopy, 153:* 393–403.

Weibel, E. R. (1992). Stereology in perspective: A mature science evolves. *ACTA Stereologica, 11,* Supp. 1: 1–13.

Weibel, E. R., Staubli, W., Gnagi, H. R., and Hess, F. A. (1969). Correlated morphometric and biochemical studies on the liver cell. Morphometric model, stereologic methods and normal morphometric data for rat liver. *Journal of Cell Biology, 42:* 68–91.

Weis, S., Weber, G., and Rett, A. (1992). A stereological investigation on magnetic resonance imaging scans in Down syndrome. *ACTA Stereologica, 11,* Supp. 1: 175–180.

West, M. J. (1990). Stereological studies of the hippocampus: A comparison of the hippocampal subdivisions of diverse species including hedgehogs, laboratory rodents, wild mice and men. In J. Storm-Mathisen, J. Zimmer, and O. P. Ottersen, eds., *Progress in Brain Research.* Elsevier, Amsterdam, pp. 13–36.

West, M. J. (1993). New stereological methods for counting neurons. *Neurobiology of Aging, 14:* 275–285.

West, M. J. (1999). Stereological methods for estimating the total number of neurons and synapses: Issues of precision and bias. *Trends in Neuroscience, 22:* 51–61.

West, M. J., and Gundersen, H. J. G. (1990). Unbiased stereological estimation of the number of neurons in the human hippocampus. *Journal of Comparative Neurology, 296:* 1–22.

West, M. J., Slomianka, L., and Gundersen, H. J. G. (1991). Unbiased stereological estimation of the total number of neurons in the subdivisions of the rat hippocampus using the optical fractionator. *Anatomical Record, 231:* 482–497.

Wickelgren, I. (1996). Is hippocampal cell death a myth? *Science, 271:* 1229–1230.

Wicksell, S. D. (1925). The corpuscle problem. A mathematical study of a biometric problem. *Biometrika, 17:* 84–99.

Wicksell, S. D. (1926). The corpuscle problem. Second memoir. Case of ellipsoidal corpuscles. *Biometrika, 18:* 151–172.

Williams, R. W., and Rakic, P. (1988). Three-dimensional counting: An accurate and direct method to estimate numbers of cells in sectioned material. *Journal of Comparative Neurology, 278:* 344–352.

Index